STEM Career Dynamics

A Comprehensive Guide to Getting a Job in Science and Technology

ALSO FROM COLD SPRING HARBOR LABORATORY PRESS

Lab Dynamics: Management and Leadership Skills for Scientists, Third Edition

At the Helm: Leading Your Laboratory, Second Edition

Career Opportunities in Biotechnology and Drug Development

STEM Career Dynamics

A Comprehensive Guide to Getting a Job in Science and Technology

Carl M. Cohen, Ph.D.

COLD SPRING HARBOR LABORATORY PRESS
Cold Spring Harbor, New York • www.cshlpress.org

STEM Career Dynamics: A Comprehensive Guide to Getting a Job in Science and Technology

Published by Cold Spring Harbor Laboratory Press, Cold Spring Harbor, New York

Publisher	David Crotty
Acquisition Editor	John Inglis
Project Manager	Barbara Acosta
Editorial Assistant	Danett Gil
Permissions Coordinator	Carol Brown
Production Manager	Denise Weiss
Cover Designer	Mike Albano

Front cover: Newton's cradle. Modified from iStock image 1071025122; digitally created by fdmsd8yea, Nerima-ku, Japan.

ISBN 978-1-621825-55-5 (hardcover)
ISBN 978-1-621825-56-2 (epub)
LCCN 2026937491

Contents

Acknowledgments

Writing this book was a family matter. I am blessed by having a wife and adult children who are my brain trust. As a group they can be relied upon to provide sage advice, helpful criticism, and reliable alerts when I get too full of myself. As individuals they are each in their own way my guiding lights. My wife Suzanne is my partner in life and in our business, Science Management Associates, and she coauthored the book *Lab Dynamics: Management and Leadership Skills for Scientists* with me. Suzanne had the patience and fortitude to read every page of this book more times than she or I can recall. Thankfully, her feedback to me on the early versions was informed by her years as a psychotherapist—seeking to help rather than judge. On the occasions when I saw her reading early chapters, the grimaces, raised eyebrows, and furious note scribbling in the margins betrayed the degree to which she exercised self-control with me. As the book progressed, her demeanor settled to the point of acceptance and, on occasion, praise. This book would never have happened were it not for her.

My daughters, Zoe Cohen and Phoebe Cohen, also provided invaluable input. Phoebe, an accomplished scientist and educator, agreed, somewhat impulsively, on a family vacation to read and comment on the book. Her comments, editorial contributions, and insights into the challenges faced by women in science are embedded in most of the chapters. Also, because of her, numerous errors of fact on my part thankfully never made it into print. On that same family vacation, Zoe, a visual artist and labor union organizer (yes, she is unique!), quizzed me on what I was including in the book about nontenure-track or adjunct faculty positions. Her question stopped me cold. Seeing her poor dad at such a loss, she offered, and in short order produced, an outline of what precisely I needed to say. It made the book better, and as always when I talk to Zoe, I learned a lot in the process.

My sincere thanks go to Juan Matute, Serena Tucci, and Brett Abrahams for generously sharing their own experiences and personal anecdotes regarding academic and industry life in general and the job market in particular. Thanks also to Alejandra Ochoa Escheveria, for her generous input and numerous conversations about the academic job market.

Thanks also to John Inglis and the entire staff of Cold Spring Harbor Laboratory Press for their decades-long support of my work and especially of this book. Thanks to the production team at KnowledgeWorks Global as well for their expert ability to turn a manuscript into a finished product.

Finally, I want to thank all of my coaching clients, current and past, and the many participants in my workshops over the years. Hearing your stories and concerns alerted me to the need for this book and helped shape the choice of topics covered.

Carl M. Cohen

Introduction

Welcome. If you are reading this, it is likely that you are looking for a job in STEM. My goal with this book is to make your search process more efficient, productive, and ultimately successful.

As I did the background research for this book, I found many resources online and in print providing guidance on specific aspects of seeking a job in STEM but no one place or resource where the best ideas and approaches were compiled. This was especially true for jobs in the academic sector, where getting a job is "an opaque process with no clear standards or guidelines."[1] Job candidates in STEM spend months sifting through online resources offering conflicting and idiosyncratic views of how to go about getting the job they want. This book is a one-stop shopping resource for those candidates.

STEM workers periodically face incredibly competitive job markets, with a crush of candidates having the scientific background, experience, and qualifications that are close to or even identical to what any given position requires.[2] At such times, you need to distinguish yourself from the dozens or hundreds of other technically qualified applicants applying for the same job as you are. One way to do that is to present and highlight your scientific skills in a way that distinguishes you from the pack and attracts employers' interest. This book will show you how to do that.

But scientific skills are not enough in today's competitive job market. You also need to demonstrate or acquire key *extra*scientific skills that set you apart. These skills include managerial, supervisory, and leadership skills, as well as interpersonal skills involved in giving and receiving feedback, negotiating, team building, and mentoring. People with these skills typically also have high emotional intelligence and are empathic, introspective, and self-controlled.

When I hired scientists, these skills were always the "tie breakers" that would help me choose between two otherwise identically qualified candidates. Hiring someone with good interpersonal and self-management skills in addition to their scientific skills meant that when difficult or contentious issues arose, as they do routinely in science, I could trust these people to work through most of them in a level-headed manner and with sensitivity. It also meant that when I needed to find someone to supervise another scientist or to move up into a vacated managerial role, I would have ready-made candidates waiting in the wings. It meant I would not need to do a new external search, saving me time and money.

Developing these extrascientific skills is the topic of *Lab Dynamics: Management and Leadership Skills for Scientists*, 3rd ed. (Cold Spring Harbor Laboratory Press), which I will be referring to as a resource throughout.[3] *Lab Dynamics* provides practical advice on how to succeed in science and technology and has sold over twelve thousand copies since the first edition was published in 2005.

If you already have the skills presented in *Lab Dynamics*, this book will show you how to make that apparent in a way that can make the difference between getting a job offer and getting ignored. If you think you need to be better at these skills, use *Lab Dynamics* as the instruction manual to help you gain them.

This book's fifteen chapters show you how to display your scientific and extrascientific skills and experience in the best possible light.

Chapter 1. The Great Scientific Disruption: Why Your Career Needs a Strategy (Not Just Good Science) describes the cyclical and unpredictable nature of the STEM job market, the importance of extrascientific interpersonal skills on getting a job, and the crucial impact of AI on the job application process. **Chapter 2. The Connection Equation: How to Network Your Way Out of Invisibility** will help you lay the groundwork for your job search by showing you how to build and maintain a network of professional and personal connections. It also provides guidance and scripts on how to use your network when you need it most—in your job search. **Chapter 3. It's Not About You: Cracking the Code of Academic and Industry Job Hunting** shows you how to treat your search as a project, starting with a realistic timeline. It also shows you how to decode job postings and customize your résumé or CV to address an employer's needs. Whether you are staying in academia, or moving into industry, **Chapter 4. The Craft of Cover Letters: Industry vs. Academia (and Why AI Won't Save You)** introduces you to the all-important cover letter, how and why you should customize your letter for every job you apply for, and the key differences between academic and industry letters. Chapter 4 also shows how you can use AI to help and how an overreliance on AI can backfire on you.

Chapter 5. The Academic Trifecta: Research Vision, Teaching Philosophy, and Strategic References introduces two key elements of academic applications, the research and teaching statements, with step-by-step instructions on how to prepare them. It also guides you in selecting your best references and shows how you can help them help you. **Chapter 6. Dancing on Eggshells: The Guide to High-Stakes Academic and Industry Interviews** covers what for many is the most stressful part of the job search process. It shows why extensive research and preparation are crucial for both screening interviews and full on-site visits and covers practical aspects of phone and video interviews. It also provides extensive lists of common interview questions for you to practice. **Chapter 7. A Deeper Dive: Acing Behavioral Interview Questions** prepares you for being grilled by sophisticated interviewers who may ask questions you are not expecting and gives you options for how to respond. This chapter also provides lists of questions that you need to ask your interviewers both to understand the job and to show that you have prepared for the interview.

One of the most important parts of the job interview process, giving presentations, is covered in **Chapter 8. Showtime: The Three Performances That Can Make or Break Your Career**. This chapter contains detailed instructions on how to give successful job talks, chalk talks, and teaching demonstrations. It covers in detail the crucial distinction between a research talk and a job talk. It also has extensive practical guidance on preparation and delivery of your talk, how to handle tough questions, and how to avoid the most common mistakes for each type of presentation.

Once you are successful in your interviews and presentations and get an offer, Chapters 9, 10, and 11 give you the tools you need to negotiate terms. **Chapter 9. The Negotiation Playbook: Data, Tools, and Tactics for Job Offer Success** gives you a comprehensive toolkit for conducting a successful negotiation on the terms of your employment. Included are guidance on preparation strategies such as defining your ideal, acceptable, and bottom-line outcomes, focusing on shared interests rather than positions, and using standards and criteria to justify requests. The chapter also covers specific strategies for different negotiation platforms including video, phone, and in person, as well as offering guidance specifically for women. **Chapter 10. The Art of the Academic Deal: Negotiating Your Worth Without Burning Bridges** is a step-by-step guide to the complexities of negotiating an academic job offer, including all of the terms that are negotiable, presenting and defending your startup package, and how to negotiate teaching and other academic commitments. **Chapter 11. Simpler but Not Easier: The Strategic Guide to Industry Job Negotiations** shows the sequence of steps involved in a typical industry job offer and covers each of the major negotiable elements including salary, bonuses, and stock options.

Chapter 12. From Scientist to Science Leader: Landing Management and Leadership Positions shows scientists seeking to transition into management and leadership roles how to decipher management job postings, how to identify the skills specified for the job, and how to match those with their own skills. It also covers how to demonstrate core leadership competencies in applications and interviews. **Chapter 13. Meet the CEO** outlines how to excel in the all-important interview with the CEO. It shows how to anticipate common questions asked by CEOs, how to show authenticity and resilience, and how to showcase leadership skills, ask insightful questions, and negotiate compensation and role terms.

In **Chapter 14. When to Walk Away: Red Flags and Warning Signs**, you will learn how you can be sure that a job opportunity is the right fit for you before you accept the offer. The chapter outlines twelve key warning signs that could prompt you to reconsider an offer and provides guidance on when to gracefully withdraw from the application process if necessary. Finally, **Chapter 15. Getting to No: Understanding Rejections and Declining Offers** will help you understand and manage your reactions if they decide you are not the right candidate. It will also show you how, and how not, to let them know if this position is not the right one for you.

After reading this book you will be ready to take control of your career by doing the following:

- Creating a network of contacts to generate opportunities
- Understanding what employers are looking for
- Presenting yourself as the solution to their needs

- Negotiating your best possible offer
- Making decisions based on data

Using the tools and techniques presented, you will learn how to put your best foot forward and get the STEM job you want and deserve even in the most competitive job markets.

How AI was used in this book

In all places in this book where I used an AI tool, I have indicated so explicitly in the text. Most of the AI input came from Claude, but ChatGPT was also used on occasion. AI was used in the creation of first-draft chapter summaries at the end of each chapter, and these were then further edited by me. It was used in the generation of the commentary on the industry cover letter in Appendix 1 of Chapter 4. In some cases, AI was used to suggest or generate chapter or section headings. In all cases, AI-generated content was heavily edited. Other than this, all content of the book except that which I have quoted or attributed to others was written by me.

1 The Great Scientific Disruption

Why Your Career Needs a Strategy (Not Just Good Science)

- ▸ Cycling through the STEM job market
- ▸ Basic or applied?
- ▸ "But I just want to do science"
- ▸ Is a B.A. or B.S. enough?
- ▸ Is doing a postdoc worth it?
- ▸ The impact of AI on your application process
- ▸ Chapter 1 summary
- ▸ Appendix 1. What to expect when moving from academia to industry

CYCLING THROUGH THE STEM JOB MARKET

Getting your first or next STEM job can be the most satisfying or the most frustrating professional endeavor you will ever engage in. Granted, if you are a theoretical astrophysicist, your job prospects were never going to be much better than winning the Mega Millions lottery, no matter what the decade. But if you were in the right field at the right time, say computer science or coding between 2000 and 2020 or biotechnology in the early 2000s, you would have had to beat the recruiters away with a stick. Between 2002 and 2012, Internet sector and software-related jobs increased by 634% and 562%, respectively, and life science–related jobs by 300%.[4] But if you were in those fields then and you are still around now, you also know that good times do not last. Scientists in these fields have seen formerly in-demand jobs suddenly become scarce, as technical advances (think AI's impact on coding jobs),[5] market forces (availability of venture funds for biotech startups),[6] or dramatic changes in US government funding of science (2025!) bring their value or viability into question.

Such changes periodically reshape entire fields, leaving scientists who never had to break a sweat to get a job interview suddenly scurrying even to get their résumés read. If you are in or planning to go into science, you need to accept this as a feature, rather than a bug of the world you have chosen. Even those who have not yet been scarred by loss of employment can still face disillusionment-inducing wake-up calls. While you went into science because you wanted nothing more than to unravel the secrets of the natural world, to bask in the beauty of a mathematical proof, or to improve the lives and health of mankind, you may find your aspirational bubble burst in a variety of ways. Finding yourself working for people whose idea of a lofty goal is to increase their Q4 earnings is just one of many.

If you thought you could avoid such short-term, narrow-minded thinking by going the academic path, you will find other barriers to soul-satisfying work. Generating a robust stream of scholarly publications, feeding the ravenous maw of laboratory expenses by applying for increasingly competitive research grants, and keeping your eye on the ticking of the tenure clock end up consuming more hours than seem to exist.

And if that were not bad enough, along came 2025, which will undoubtedly go down as the year in which the greatest transformation ever seen in American science took place. The radical restructuring of priorities by the US government has created turmoil in the US scientific ecosystem that will not be easily remedied. The wholesale dismissal of federal scientific workers in all disciplines and at all levels—from research assistants to Directors of the National Institutes of Health (NIH)—along with the cancellation, or threats thereof, of governmental grants and funding was unprecedented and catastrophic. Whether 2025 will go down in history as a scientific mass extinction event or just a short-term but nonetheless devastating error remains to be seen. What is clear is that the impact on tens of thousands of working scientists was immediate, shocking, and unexpected.

Such turmoil may dissuade some from entering science as a profession and motivate current scientists to seek greener pastures. A 2025 *Nature* survey found that 75% of US-based scientists answered "yes" to the question "Are you a US researcher who is considering leaving the country following the disruptions to science promoted by the Trump administration?"[7] But surveys like this one can attract more responses from those who agree with the premise of the survey than those who do not, so it is hard to know how representative the 75% figure is. I believe that in the long term, despite periodic sea changes in funding and scientific priorities, the United States has been and will continue to be the best place in the world to do great science.[8] Even in times of reduced funding and job scarcity, there will be turnover in some areas and new opportunities in others. When the competition for positions is great, you need to put your best foot forward so you can get not just *any* job but the job you *want*. This book will show you how.

BASIC OR APPLIED?

The first question you need to answer as you begin your job search is "What scientific field am I in?" This should be an easy one, but do not think that it is carved in stone. It is not unusual for scientists to migrate from one discipline to another over the course of their career.

The second question is whether you want to apply existing scientific knowledge toward making new things (applied science) or create new scientific knowledge that extends what we know about the natural world (basic science). This question is as relevant to those seeking their next job as it is to those seeking their first. And as with the first question, the answer is not carved in stone. Scientists routinely move from basic science in academia to applied science in industry, and even sometimes the other way.

To help us dissect career paths, the figure below shows scientific disciplines arranged on a line with engineering at one end, discovery science at the other, and what we will call applied science in the middle. An engineer attending one of my workshops told me that he had never tested a hypothesis. What he meant was that engineers apply proven laws and principles to new problems in order to design and create new things. No engineer wants to use a novel or untested theory to design the world's tallest building, and

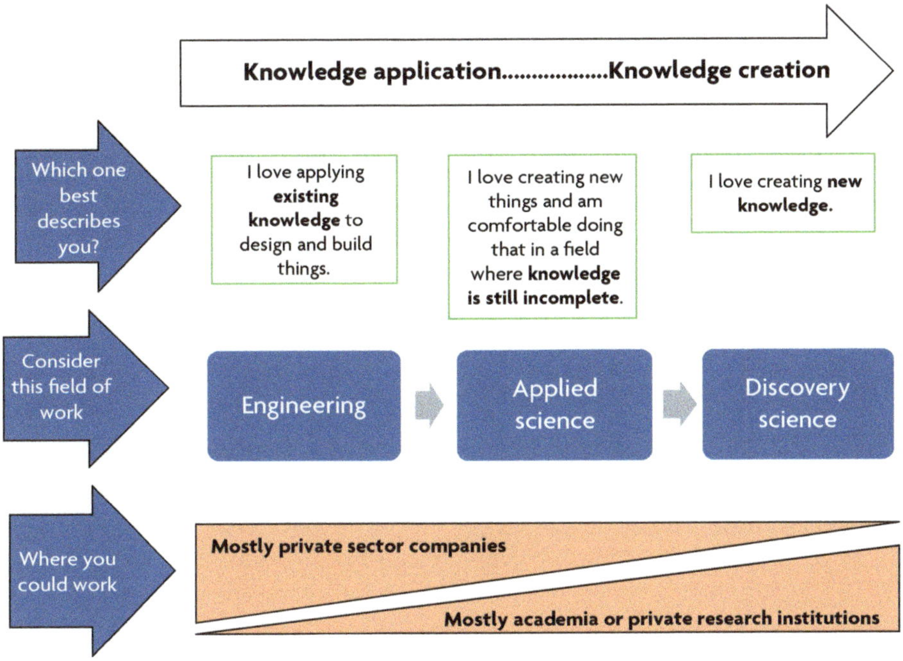

even if they did, the potential occupants would surely object. Even though engineering uses principles that are known and tested, engineering can be and is a highly creative endeavor. The dizzying array of today's newest technologies, instruments, and products attests to the immensely creative minds of the best engineers. The wild success of Apple products, for example, shows that engineers routinely create objects of great beauty and functionality.

Next comes applied science. Life scientists working in biotech in the pharmaceutical industry, and chemists and physicists working in the semiconductor industry, are all doing applied science. They apply what they know to find solutions for things that they do not know. In applied science, you know only some things because you are working in scientific fields where the rules are mostly but not entirely understood. Because of that, hypotheses can be tested, and new scientific discoveries can be made along the way. Like engineers, applied scientists work mostly in the private sector and typically focus only on company objectives.

What if you aspire to be more than just a creator of new objects? What if what gets you out of bed in the morning is a deep desire to understand the universe and to apply your curiosity to uncovering aspects of the natural world that we still do not understand? In that case, discovery science is for you, and your job search will almost certainly be focused on academic or private research institutions.

The figure also shows that while most engineers will end up in industry[a] or private sector companies, some will land in an academic setting teaching and pursuing

---

[a] Throughout the book, I will use the word "industry" to refer to all nonacademic for-profit employers.

cutting-edge technologies and applications. Most discovery scientists will end up in academia or private research institutions (similar to academia in terms of freedom to pursue independent research but without teaching and tenure), but some will find themselves in (increasingly rare) companies that have basic research programs. Others will work in government agencies, such as the NIH, National Aeronautics and Space Administration (NASA), and Department of Defense (DOD).

As a job applicant, you need to understand that industry or academia have dramatically different expectations for prospective employees. Some of the most common differences are listed in Appendix 1. This is the most important difference between an academic faculty position and an industry position:

> *In academia they are hiring you to work on your projects; in industry they are hiring you to work on their projects.*

This applies *only* to academic faculty positions, not postdoctoral positions. The highest-level academics are tenure-track faculty who function like mini-CEOs of their own small companies. This is the quintessential job if you want to be in the business of discovery or knowledge creation, and such people are often called principal investigators, or PIs, the term used by the NIH and the National Science Foundation (NSF) to denote the lead person of a research grant.

Lower on the academic ladder are research scientists and postdoctoral fellows (postdocs) or trainees who work for a PI in their lab. Such positions can contain elements of both academic and industry roles. Like an academic, you will be working on making fundamental discoveries, but like in industry, and this is important, you will be working on your PI's projects, not your own.

If you are applying for a postdoc position, you may get lucky and have an adviser who understands that an important part of their role is mentoring. With such an adviser, you will be in a hybrid role, working on some projects of their devising and possibly others of your own. In the best cases, they will take their role as mentor seriously and offer you regular professional development and career guidance and will be helpful when it comes time for you to look for your next job. Indeed, as we discuss later in this chapter, and also in Chapter 14, the reputation of your potential adviser as a mentor should be a key determinant of whether you choose to apply to their lab as a postdoc.

With increasing frequency, postdoc positions can morph into the role of research associate or research scientist, which in some cases can last for years. For those who love doing research, get on well with their boss, and have little desire to write grant applications, such positions can be extremely gratifying. These positions typically offer few opportunities for advancement or independence, and for many, that is just fine. For others, such positions are viewed as interim roles to occupy while they try to create a body of work that helps them gain independence. A generous boss who allows them a degree of independence is the key for this strategy to succeed. A less generous boss who views their associate as a pair of hands to advance only their own agenda can make that strategy difficult or impossible.

Being able to work on independent or semi-independent projects almost never happens in industry positions. A failure to appreciate this facet of industry can cause naive job applicants great disappointment. When applying to industry, you need to convince them that you plan to be laser focused on applying your skills to their projects. Period.

"BUT I JUST WANT TO DO SCIENCE"

If your deepest desire is to continue doing hands-on science throughout your career, you will find both academic and industry tracks challenging. In industry, as you advance in seniority and accomplishment, you will more and more be pushed into a managerial role. This is partly for practical reasons—promotion to managerial roles is almost always the only way you can advance professionally, gain responsibility and some independence, and get paid more. Few companies have what might be called pure science tracks, where scientists advance in title and compensation without taking on any managerial responsibilities.

This is also true for academia, except, as a faculty member, you will need to take on supervisory responsibilities for graduate students, postdocs, and technicians almost from day one. Apart from not having graduate students, the same is true even if you have an appointment as a research faculty member. In this nontenure-track role, you have your own research and salary funding from external agencies and do your own research; you are essentially renting lab and office space from the institution. Research faculty positions typically come with no institutional responsibilities other than obeying whatever rules and regulations apply to being an employee of the institution. Such positions are attractive to scientists who are able to secure research funding and do not want to deal with the responsibilities of being a tenure-track faculty member, such as teaching and committee work.

Research faculty often have the same titles (assistant professor, etc.) as tenure-track faculty and can advance in rank (from assistant to associate to full professor), as do tenure-track faculty, but without the possibility of tenure. Rules for research faculty vary widely between institutions, including what support, if any, they can receive if their research funding is lost. For tenure-track faculty, there may be an expectation that the institution will provide bridge funding between grants. A high profile or consistently well-funded research faculty member may be able to negotiate such terms, but they are not a given (see Chapter 10). It may be possible in times of lean funding for a research faculty member to supplement their funding by teaching or other forms of service. If you do take a nontenure-track or adjunct position, you should not assume that this will give you an inside track on even getting considered for a tenure-track position when one becomes available.[9] Such positions are filled via national research, and your candidacy will likely be treated just like anyone else's.

Adjunct faculty are also nontenure-track positions and are most often part-time faculty hired to teach one or more courses; they typically have no contract or long-term commitment from the institution. Adjunct teaching faculty positions have become increasingly attractive to institutions, in part because they can pay adjuncts far less than tenure-track faculty and do not have to pay them benefits. Adjunct positions are typically focused on teaching, so many of the myriad hurdles required to obtain a tenure-track or permanent faculty position that we will focus on do not apply to these positions.

While you will be managing people either on day one in academia or as you advance in industry, there is a big difference in expectations between the two. In industry, you will actually need to be a good manager or supervisor, not just an adequate one. In academia, much to the dismay of nearly every grad student and postdoc on earth, there is hardly any oversight or evaluation of your managerial, supervisory, or mentoring capabilities unless your actions rise to the level of being egregious or illegal.[3] The result is an alarming lack of managerial and supervisory skills on the part of many PIs, which almost certainly is a contributory factor to the epidemic of depression and anxiety among science trainees.[10,11] Based on my own firsthand observations as a coach of academic, industry, and government scientists at all levels, those with poor managerial and mentoring skills eventually pay the price in the form of higher rates of lab personnel problems and conflicts, higher turnover, and, ultimately, reduced productivity.

So, don't think that taking a job in science means that all you will be doing is science. For most scientists by far, the most challenging part of their job is not the science but the people. As a scientist with a penchant for dealing with data, concepts, and objects, learning these people skills may be harder than learning quantum electrodynamics or gene editing. Managing, supervising, mentoring, and collaborating are skills that need to be learned by every practitioner of science in every field of endeavor. Unfortunately, you have likely never learned them at any point in your scientific education, and unless you end up at a large company with a formal management training program, you will have to acquire them on your own.[12] For that I refer you to *Lab Dynamics*.[3]

There is another important reason to acquire skills in the interpersonal and managerial realm: At least for the foreseeable future, it will be harder for AI to replace people with those skills. A recent *Wall Street Journal* article opined that people who developed capabilities to work collaboratively and across disciplines have "skills [that] remain highly valued to employers, and may never be automated away."[5]

IS A B.A. OR B.S. ENOUGH?

If you are an undergraduate in a STEM field, you have several options after you graduate. In certain fields like computer science and engineering, you could move right into an actual job. There is a lot to be said for getting to work right away rather than going on to an advanced degree—you start earning money and paying off your student loans right away. But the availability of entry-level jobs for college grads changes cyclically and in response to technological advances like AI.[13] Getting a higher degree will increase your job possibilities as well as potential for future growth and earning potential.

In other fields such as biology, physics, or chemistry, the only job you are likely to get with a bachelor's degree is that of a lab assistant or technician. If you aspire to do more, you are going to need a graduate degree, i.e., an M.S. or a Ph.D. If you are unsure whether that is the right path for you, take an entry-level job and see how you like the work. The hands-on experience you gain can help you clarify your goals and get into a good graduate program if you decide on that path.

IS DOING A POSTDOC WORTH IT?

Many who go on to get an advanced degree such as a Ph.D. continue their training even further by taking a postdoctoral position. That can provide advanced training and experience in a field different from the one they got their Ph.D. in, allow them to add to their publication list, and lay the groundwork for establishing an independent research career. If you aspire to the academic tenure track, a postdoc is almost a de facto requirement. In these positions, those who did postdocs outnumber those who did not by 2.5-fold.[14]

If your goal is to work in industry, doing a postdoc may not offer any advantage. From a financial perspective, your best bet as a recent Ph.D. recipient is to get to work right away.[14,15] Nonetheless, about half of the Ph.D.s who go into industry have done a postdoc.[14]

This book does not have a separate chapter on applying and interviewing for postdoctoral jobs. With one or two caveats, the guidance provided in the chapters on industry is what you should be following. This is because in most cases as a postdoc you will be expected to work on projects that are the focus of the lab you will be joining, not your own projects, similar to what you should expect in industry. What this means is that during the application and interview process discussed later in the book, you should focus on conveying your enthusiasm and interest in what the lab works on. But, unlike in industry, postdoc positions typically have an element of training to them; you might not be expected to have all or most of the skills you need on day one. As with scientific funding and jobs in general, the availability of postdoc positions is influenced by economic and political boom and bust cycles, with 2025 being notable for dramatic losses.[16] If you do decide to go the postdoc route, choose the lab you will work in carefully. Many postdocs erroneously assume that their best choice is a lab run by a famous scientist, a Noble Prize winner, for example. In fact, the most important factors in your choice should be the following: the lab's focus matches your interests, their current productivity (in the form of publications), and, *just as important as the first two combined*, the reputation of your adviser as a mentor and as an advocate for you when it comes time to find your next job. Without this last feature, you may find yourself adrift or scurrying to find a job when your three-year stint is completed. When interviewing for postdoc positions, invest time in talking one-on-one with the PI's current or recent postdocs to find out how helpful or unhelpful this person has been. If you have a choice of where to go, this factor should be the tiebreaker. See Chapter 14 for a list of questions to ask current or past postdocs of the PI you are considering working for.

THE IMPACT OF AI ON YOUR APPLICATION PROCESS

Throughout this book, I will refer to the availability and use of various AI tools to help you with the application process. Given the ubiquity and utility of these tools, you would be remiss not to use them. Because of this, you will get no credit for having your cover letter, résumé, or CV perfectly formatted and well written. If, on the other hand, they are poorly written, contain misspellings, and are in the wrong format, your application may be rejected right away as the reader may conclude that you are an unskilled writer, did not pay attention to instructions, are unfamiliar with AI, or all three.

Because AI makes it so easy to feed in instructions and raw data and get out cogent, clearly written documents, there is no longer any reason for all written materials you submit as part of an application not to be perfectly written. Of course, you still need to feed the AI agent your information, ideas, plans, etc., in order for it to create your document. Asking an AI agent to generate ideas for your research plan or to actually create the plan for you may sound like a great idea until you read what comes back: bland, generic writing and references to research articles that do not exist. Feeding *your* research plan to an AI agent and asking it to organize and format it is a better idea, but even then, you need to spend a lot of time checking what it wrote because some if it will be lunacy and hallucinations.

Because employers can no longer rely on anything submitted in written form to gauge organizing and writing skills, AI is fundamentally changing how employers screen and interview applicants. There will be an increased emphasis on face-to-face interactions during the screening and interview process, including more and more employers asking questions designed to assess skills in specific technical areas.[17] It is impossible to pretend you know how to use the genetic tool Cre-loxP to make an inducible knock out of an immunoglobin gene in a mouse and what such a mouse could be used for if you are sitting right there in front of the people asking the questions, instead of in front of your laptop.

So, while AI may make your life as an applicant easier in some respects, the increased in-person scrutiny you may receive could make it harder. This will be especially true for entry-level jobs in science. Applicants for entry-level jobs mostly do not have a list of publications that document what they have done or a prepared seminar of their original work. All employers have to go on is the applicant's word or the word of their references, who may not be 100% forthcoming. Without such supporting evidence, asking hypothetical or problem-solving questions will become more important as screening tools.

Using AI to organize or format documents is fine, but do not try to use it to fool hiring managers into thinking that you are someone you are not. In science, if you are not who you say you are, cannot do what you say you can do, or have not done what you say you've done, you will flame out sooner or later. That does not mean you should not work to portray your skills and experience in the most positive light. Scientists too often fail to get the job they want not because they lack the skills needed but because they fail to highlight properly the skills they do have. You will learn how to do that here.

CHAPTER 1 SUMMARY

STEM job market dynamics

- STEM job markets are cyclical and unpredictable.
- Previously high-demand fields can suddenly become scarce because of technological advances, market forces, or policy changes.
- 2025 is highlighted as particularly disruptive for American science owing to government restructuring.

Types of scientific work

- Scientific jobs exist on a continuum from engineering (applying proven principles) to applied science (working in partially understood fields) to discovery science (basic research).
- Key distinction: In academia, you work on your own projects, while in industry, you work on theirs.

Career progression and management

- Most scientific jobs eventually require scientists to take on managerial responsibilities.
- Pure hands-on science roles become increasingly rare as people advance in their career.

Impact of AI on job applications

- As AI makes it easier to create polished application materials, employers will rely more heavily on face-to-face interactions and technical questioning during interviews to assess candidates' actual skills.
- This shift requires candidates to be prepared for a more rigorous in-person evaluation.
- Do not use AI to pretend to be someone you are not.

Keys to success in science careers

- Technical competence alone is insufficient; strong people management skills are essential.
- The ability to present your skills, capabilities, and accomplishments in a positive light is crucial.
- Having realistic expectations about the challenges and cyclical nature of scientific employment will help you be more prepared for and anticipate change.

APPENDIX 1

What to expect when moving from academia to industry[18]

If you are considering a job in industry, you will need to adjust your expectations of what doing science will be like. While science in academia is most often focused on discovery and the pursuit of knowledge, in industry the focus is on pursuing a product and profit. Working in industry in pursuit of a product can be every bit as exciting and satisfying as working in academia in pursuit of knowledge, but the sooner you understand and accept the cultural and economic differences between these two domains, the happier you will be. Below are key differences between these two work domains. It is better to be aware of them on day one than to have your illusions eroded slowly over years.

1. **Your job is to advance projects *according to a schedule and within a budget.*** Consequently, you may be asked to stay focused in a way that is different from what you were used to in an academic setting.
2. **Many projects may seem scientifically fascinating, but the only science that matters is what will generate revenue and profit one day.**
3. **Your funds come from investors who need to see evidence of regular progress to ensure that they have invested wisely.** Consequently, if progress on a project is slow or nonexistent, your company may decide that it is no longer worth pursuing.
4. **Your project may get canceled.** There may come a time when a project that you are working on gets canceled or is set aside for reasons that you do not understand. It may look to you like things are moving along nicely, but to others in management this same project may look too risky or lack sufficient results to support its continuation. Ideally, especially in larger, more established organizations, your company will have other projects in the pipeline that you will get assigned to. Bigger, more established companies pursue multiple projects and approaches simultaneously. If one is not moving fast enough or another is looking especially promising, they may shift resources (meaning you) as needed. This is another difference from academia. The work you do needs to be work that the company needs to get done at that moment, so you may be called upon to support their most promising efforts. In smaller, leaner start-ups where there may be only one or two major foci, this will not be the case, and you may find yourself looking for another job.
5. **You will be a member of a team.** In academia, many projects are one-person operations where you have a high degree of ownership and control over what gets done and in which you get personal credit for its success. Companies leverage the power of teams to speed up progress. Your results and work output will support your team's mission, as opposed to that of a personal project. Your immediate manager will review and guide the work you do and will be the primary conduit of recognition for the work you do as well as for guidance on improvement if needed.
6. **It is not your data.** Generating exciting data is exhilarating, and you will justifiably want to be recognized for the work you do. Ideally, you will have opportunities to

present and get credit for your work in various ways. But the data is not yours; it belongs to the company. Hoarding data or keeping it secret so you can use it for your personal benefit is not an acceptable way to highlight your skills.

7. **Teams have multiple levels of oversight and input.** In academia, your lab head was likely the only person to whom you were responsible. In industry, there are multiple layers of managers to whom you are either directly or indirectly responsible. Your work will be managed by your team leader, who in turn is managed by a supervisor, director, or VP, all of whom are responsible for ensuring that the work you and your team do is focused and productive. To academics, it can seem confusing when someone from, for example, business development or even finance asks questions about a project that you are working on. In industry, everyone has a stake and may have input in what you are doing. When evaluating the utility of a new project or the viability of an existing one, many considerations beyond science come into play. Could this project lead to a viable product? If so, how long might that take? Can they find a corporate partner to support it? Do they or others have intellectual property associated with the project? What is the anticipated market for this product?
8. **Welcome to matrix management.** If you become a member of a "functional team," the management structure may be unfamiliar to you. If, for example, you are in the Assay Development Group, you will report to your team leader or manager (often called a "line manager"), who is responsible for ensuring that you perform your work up to scientific expectations. Your actual working time, however, may be fractionally allocated to multiple projects, each managed by a different project manager who is responsible for progress on that project. So, you now have two kinds of managers: Your project manager is responsible for what you do and your line manager for how you do it.
9. **Love the science, but do not fall in love with it.** Academic projects often spawn offshoot projects leading to new directions in research. In a company, this may not be possible. It needs to stay focused on achieving its objectives.
10. **Always keep in mind the difference between "perfection" and "good enough."** Perfection is fine if it does not take forever, and it typically does. Consequently, you may be asked to come up with solutions that are good enough to solve the problem so that you can move on to the next problem. Good enough today is typically better than perfect tomorrow.
11. **Management may talk about your science in ways that you are not used to.** As a scientist, you are trained to be conservative in how you talk about your results, weighing all possible interpretations and stating conclusions as being tentative and requiring multiple replications and verifications. You may find that when management talks about these same results they express an optimism that makes you uncomfortable. You see your results as "preliminary," "tentative," and "needing replication," and they talk about them as "exciting," "groundbreaking," and "proof of concept." This is all OK and is an example of you doing your job and them doing theirs.

12. **Power structures and hierarchies are facts of life.** For some decisions, you may be asked for your input while for others you may not. Spirited debate about the interpretation of data and the next steps to take are a hallmark of good science, but on occasion, after all the debate, a decision may be made that you do not agree with. You need to accept that, once you have given your input, it may be someone else's responsibility to make the decision. This approach enables companies to make decisions in a nimble manner and adhere to their timelines.

2 The Connection Equation

How to Network Your Way Out of Invisibility

WHY NETWORK?

Some scientists lead a blessed existence. When they were finishing grad school, their mentor introduced them to a colleague in their field at a scientific meeting who offered them a postdoc on the spot. During their postdoc they were extraordinarily productive and published eight high-profile papers. After three years of this they were contacted by a colleague at an elite R1 university suggesting they apply for an open faculty position that was about to be advertised.[19] They applied, were accepted, and got a big research grant, and you can fill in the rest. This is every scientist's fantasy, and for most it is just that, a fantasy.

It may seem like things just fell into this person's lap, but of course things fall into laps for a reason. This person is in the scientific 1%. They never needed to work at creating or maintaining a network because people reached out to them rather than vice versa. These people have extensive networks that are of great value to them; they just do not have to do the kind of work most of us do to create them. If you are not in that 1%, what do you need to do? The answer is networking.

Networking may be something that many of you have never even thought about spending time on. But networks—all the people you know, have ever met, worked with, talked to face-to-face, or communicated with via email—are your best source of advice, job leads, and favors of all sorts. Even if you are in the charmed 1%, there may come a time when your fortune changes—your company folds or your grant is not renewed or, worse, it is cancelled (greetings from 2025).

The Grass Is Always Greener...

Organizations are much more interested in external talent than in their own employees to fill vacancies. Top channels for quality hires:

Note: Data based on 2017 survey of 3,973 talent-acquisition decision-makers in corporate HR departments and LinkedIn members. Source: Modified from Peter Cappelli, "Your approach to hiring is all wrong," with permission from Harvard Business Publishing © 2019.[20]

At times like these when you need to pivot quickly from comfortable job complacency to anxiety-inducing job insecurity, having a good network will be your lifeline.

Connections to people in your network come in all varieties. Strong connections are those with people who know you well, while weak connections are those with casual or one-time contacts. In my case, nearly every job I ever got came via introductions from people who knew me pretty well—strong connections—and who could vouch for my credentials and skills from firsthand knowledge.

The only job I ever got by answering an advertisement was my first position after my postdoc. I responded to an ad in Science. *As it turned out, however, the hiring manager had heard me give a talk at a scientific meeting six months earlier. When he saw my application letter, he called me immediately, set up a meeting the very next day, and offered me the job on the spot. After that, every job I held in industry I got because someone I knew or had worked with had recommended me or given my name to someone. Some of these people were strong connections and some were weak connections, but they all knew me well enough and spoke about me highly enough that it was as though I came "pre-vetted." By virtue of these referrals, I was considered a known entity, significantly de-risking the hiring process.*

This is a huge benefit to the person they refer you to since you are a known entity and they do not have to do all the onerous work of ensuring that you do not have two heads. As we will see later in Chapter 3, sifting through résumés to find the gems and even selecting the top candidate from a list of three finalists is a monumentally time-consuming endeavor. The average hiring manager or search committee will jump at the chance to take a short cut and zero in on someone who has already been vetted by someone in their own organization, an old friend, a trusted colleague, or their brother-in-law. The figure above indicates that nearly 50% of corporate positions are filled via referrals.

Indeed, many companies actively encourage their employees to refer job candidates to them by offering cash bonuses as high as $5,000 to employees who refer a candidate who gets hired by the company. What this means for you is that if you know anyone at the company you are applying to, you should reach out to that person, tell them the job you're applying for, and ask them to refer you to the hiring manager or the HR department. In a field of hundreds of candidates, this may be all that is needed to propel your candidacy to the top of the list.

The same holds true for academic jobs. While the selection process for academic, especially faculty, positions may seem designed to deemphasize such nudges from within, do not rule them out. Every faculty selection process comes down to individuals on a committee or search team evaluating candidates and making convincing arguments on behalf of their choices. If one or more of these people know you, even peripherally, they may be more inclined to push for your candidacy, and do so in a forceful manner, over those who they are not familiar with.

This tendency of people to hire candidates who they see as "known entities" is especially prevalent when PIs hire postdocs. In my years of running workshops teaching scientists how to use a rational process for hiring, I have heard innumerable stories from PIs who hired a postdoc practically sight unseen, based on nothing more than the verbal recommendation of a friend or colleague. I do not recommend this approach to PIs as a hiring practice, but it is common because it is easy. The typical reason I hear such stories is because the new hire did not work out and the PI regrets their decision. The lesson for both PIs and candidates is to take advantage of networks and connections for introductions but avoid as much as possible impulse hiring decisions and impulse acceptances.

Do not think that your strongest connections are necessarily the best source of referrals. Studies suggest that weaker connections, people who you may know only slightly, are also a valuable resource.[21] The reason is that close connections operate in social and professional circles nearly identical to yours, while weaker or more peripheral connections likely operate in different circles and therefore know people you do not. So, when it comes to connections, think wide and think deep.

GET A HEAD START WITH A GOOD MENTOR

If you are a graduate student or postdoc and you are blessed with a mentor who takes their job seriously, then count your blessings. Sadly, such mentors are in short supply. If you do have one of those mentors, these are some of the things you may have already done with them:

1. Created an individual development plan (IDP[b]) that outlines your career goals and the training and skills you need to achieve those goals.

[b] An individual development plan is a tool to help you establish your career and professional priorities and to set and keep track of your learning goals. IDPs are typically done in collaboration with your PI, adviser, or mentor. IDPs are either required or recommended by certain agencies that provide funding for graduate students and postdocs, most notably the NSF and NIH. The American Association for the Advancement of Science (AAAS) provides a template for creating an IDP at www.myidp.sciencecareers.org.

2. Reviewed your IDP at least annually to make any needed changes or adjustments (people's career goals change!).
3. Received input on how realistic your career goals are, given your combination of skills, background, accomplishments, etc. This is the kind of advice most trainees never get. A mentor who gently lets you know that the jobs you are planning to apply for are probably beyond your reach is like a kind friend who tells you that you have a large piece of spinach wedged between your front teeth. They are telling you something that will be glaringly obvious to others but that you are unable to see yourself. There is nothing wrong with applying for a few jobs that are beyond your reach, but if that is all you are applying for you are going to be an unhappy camper.
4. Been introduced to others in your field, at seminars, scientific meetings, and social events. Then, by the time you are ready to take the next step, you already have a network of people who may know you by name or to whom you have been introduced, whether they remember you or not.

As noted before, in choosing a postdoctoral adviser, pay at least as much attention to the reputation of your prospective boss as a mentor as you do to their scientific field and accomplishments. You may be better off in the long run to choose an adviser who is a good scientist rather than the best in the world but who also has a reputation as an excellent mentor.

If you have a great mentor, you are off to a good start. Much of the legwork of finding and nurturing contacts will already have been started. If not, you need to start from scratch.

BUILDING YOUR NETWORK: WITHOUT A NETWORK, YOU ARE INVISIBLE

There are many good reasons to build a network, but the one people think of most often is that a wide network of contacts can be invaluable during a job search. There are really only three ways you can find out about a job. At the bottom of the food chain are jobs that you read about on a website or online portal or in a scientific journal or physical posting on an actual bulletin board on an actual wall. You know nothing about the position other than what you read there, and when you apply you become one of countless disembodied souls whose résumés are floating in Internet hyperspace.

Higher on the food chain, and rarer, are jobs that someone tells you about. If all they did was read about it in *Nature* or *Science*, it is of no more value than the first category. If on the other hand the job is in their lab, group, or company, or with someone they know, then you have taken the first step up the golden stairway to gainful employment. You may now be marginally distinguishable from the masses, especially if the recommending person knows you well (strong connection) and they know the hiring manager or anyone of importance in the organization.

Three steps in using your network in a job search

The more people you know, the more people you have kept in touch with, and the more people you update about your job search status, the greater will be the chance that you will hear about positions in this way. I know numerous scientists who have been told of positions that have yet to be advertised by their friends or acquaintances. Even if a position is advertised, some organizations will do so in obscure places that are difficult to find unless you know where to look in order to prevent a deluge of applications. Those organizations rely on their members to spread the word about open positions through their—and ideally your—network.

The final, rarest, and most valuable conduit is when a person or organization reaches out to you and asks if you would be interested in a job. This is called being recruited. It is the holy grail of job searchers (since it is not actually a search) and, as you might expect, is rarer than hens' teeth (as in, they do not have any).

If you want or need a job, looking at online postings and being receptive to suggestions from people you know are mandatory. But these approaches are passive in nature. You are basically sitting there waiting for jobs to appear in front of your face. This is a job search strategy in the same sense that planning to win the lottery is a strategy for financial independence. While there is not much you can do to increase your chances at the lottery (besides buying so many tickets that your financial ruin is guaranteed), there is plenty you can do to get out of the passive job search rut. There is only one catch: It is a lot of work. And it is the kind of work that many of us scientists do not really love doing: reaching out to people, talking to them, being curious about them and their experiences, and listening to their advice about what you should do next.

The basics of networking

Many scientists live with the unarticulated but powerful fantasy that if their science is solid and gets published in a nice journal the work will speak for itself. Someone needing exactly the kind of scientific skills in the specific technical area they have spent years working in will surely find and employ them. The truth is that they will almost never find you. But you can certainly find them.

If you come from a culture in which reaching out to others, especially those more senior to you, is not the norm, you may find this process challenging.[22] The proactive approach described in the following sections may seem aggressive or inappropriate to you, but in the United States especially this is not the case.[23] Here networking is viewed as a valuable skill that you will be rewarded for.

NETWORKING IS YOUR NEW PROJECT: TREAT IT LIKE ONE

Networking is like eating a healthy diet and exercising—benefits accrue over time without your realizing it. As a scientist you may already be familiar with some manner of networking, but you probably do not call it that. If you have ever been to a scientific meeting, presented a poster or a short talk, or gone to a social event at such meetings, congratulations, you have been networking. But only in the sense that someone flailing in the sea can be said to have been swimming. There is an art and skill to networking that transcends bumping into people at a meeting just as there is an art and skill to swimming that transcends not drowning.

Salespeople know they need to knock on dozens of doors before they make a single sale. In networking, you need to reach out to dozens and dozens of people in order to get a few good contacts. A good contact is someone who will not just open their door and talk to you but one who also is in a position to introduce you to other contacts. They are nodes in the network you are building. Each node allows you to connect to others, who themselves may be nodes. Enough nodes and eventually you have built an interconnected neural-like network of contacts.

If you are introverted, talking to strangers may be uncomfortable. But if you got this far in your scientific career, you are probably immune to a little discomfort. It is important to realize that the other person does not know you are uncomfortable. All they see is your overt behavior (looking at them, making eye contact—yes, you need to do this), and all they hear is what you say (asking them questions, briefly describing who you are). How hard is that? Networking starts with going up to someone and talking to them and then doing it again with someone else.

A TIME AND A PLACE FOR EVERYTHING, INCLUDING NETWORKING

Every time you meet someone professionally, you need to make them part of your network. Start with people you already know—your fellow graduate students, postdocs, and colleagues. Cultivating and maintaining these relationships over time will give you a solid base of contacts who will be immensely helpful to you as you progress in your career. Especially include those who may be a year or more ahead of you in graduate school or as postdocs. They will be engaging in their job search before you and can give you firsthand advice and guidance.

If you filter who you include in your network too much, you will be limiting yourself. Connections, job leads, and introduction to others come from the least expected places. The more people you have in your network, the greater your chances of making a crucial connection. This does not mean that you have to be running around at meetings and seminars like a candidate running for office, but it does mean doing more than you are probably doing now. Start small and set goals for yourself. Make a point of introducing yourself to one new person at every seminar you attend, three new people at a scientific conference, and so on. If you do not set goals for yourself, you will not do it. If you do set goals and you miss them, you can do better the next time.

ACTUAL NETWORKING EVENTS

Some times and places are just begging you to network. Social events at scientific meetings are obvious places to start. Some scientific conferences offer special badges to attendees that show they are looking for a position. At such events networking is easy because this is pretty much the express purpose of the event. You should be able to come away from one of these events having met two or three new people without trying too hard.

Cocktail hours or social events at scientific conferences are fertile sources of contacts. Especially at small conferences of less than 150 people, cocktail hour is the time you can amble up to a major figure in your field who you would never think to approach in any other situation and introduce yourself or ask a question.

Some organizations sponsor virtual networking events. For me the jury is still out on these. By definition they lack the dynamic of in-person experiences, which can be a powerful factor in connecting with a stranger. On the other hand, they are easy to attend (no travel!), and the worst that can happen is that you have spent an hour that you would have otherwise spent watching cat videos.

This story about networking was told to me by Rachel, a biotech veteran:

I had just left a position in a biotech company and was in the early stages of a job search. At this time there was an annual industry biotech conference in my city that I had attended annually as a company employee but was hesitant to attend now because it cost a lot of money to register. Nonetheless, I decided it was a worthy investment and attended. I spent most of my time wandering around the refreshment stands during the coffee breaks and before and after the scheduled networking events, looking for people I knew. One morning at a coffee break, I bumped into Sonia, who still worked at the company I had just left. We started chatting, and I reminded Sonia that I was looking for a new position, which she already knew. However, no sooner had I reminded her of this than she spotted someone on the other side of the room who she used to work with, and with whom she had spoken earlier that day. He was an executive of another biotech company in the area and had told Sonia that they were looking for someone to run part of their R&D division. Making the connection to me, she told me of this earlier conversation and offered to introduce me to this person. She introduced me to him, he introduced me to the CEO, and four weeks later I was working for them.

Commentary: The moral of this networking story is that just because you told someone a month ago that you are looking for a job, you should not expect them to remember that. It was only when Rachel reconnected with Sonia that Sonia put two and two together and thought to introduce Rachel to the executive. People need to be reminded that you exist and that you are job hunting.

Other venues include industry-specific conferences. Many local biotechnology organizations, for example, sponsor in-person get togethers or networking events for people in their geographic area. At some of these events you will end up in a room filled with people just like you—i.e., looking to make connections and maybe looking for a job. You may think you want to be in the other room, the one with the people who already have connections and jobs. But not so fast. A room full of people looking to make connections is also likely to have many people who have been doing this for a while and who can be a great source of tips and advice. It may also happen that some of them will have heard about an opportunity that is not right for them but could be right for you. The point is, everyone can be useful to you in some way, and you can be useful to them.

POSTER SESSIONS

Poster sessions at scientific meetings are great places to network. At someone else's poster, strike up a conversation with them about how fascinating their work is. If you know nothing about what they are up to, ask some dumb questions. If you know a little, engage them on some facet you would like to know more about. And if you know a lot, get into the weeds and geek out and you will have a friend for life.

If it is your poster, and people are wandering by, smile at them rather than standing there looking at your shoes. If someone wanders up, smile more, and then ask if you can answer any questions or if they work on similar things. Remember that people at poster sessions have a zillion posters they want to look at. If someone expresses an interest, do not launch onto a thirty-minute mini-seminar. Have a *thirty-second* run-through ready that covers one or two major takeaways and then stop. If they show further interest, keep going. If not, thank them for their interest and let them know you would be happy to follow up with an email with additional information if they would like. If they say yes, congratulations. You have made a new contact.

JOIN A COMMITTEE

Committees in your institution or organization are among the best places to meet people and network. If you are a graduate student or postdoc, look for committees, groups, or clubs within your institution to join. Most institutions have graduate student or postdoctoral affairs committees or other more focused committees such as clubs for those interested in going into particular technical specialties. The benefits are numerous. First, you get to meet peers from other parts of your organization. Second, you get to meet and interact with faculty, since most such groups have faculty advisers who work with them. If you are thinking that you would like at some point to advance beyond being an individual contributor or bench scientist into managing scientists, becoming an organizer or officer of such a group is a smart move. It sends the message that you have organizational and interpersonal skills. At some point, you may be able to call upon your group's faculty adviser to recommend your leadership or organizational skills based on their seeing you in action.

When I got my first faculty position, I moved from a large bustling research university to a smaller institution nearby. Despite being in the same town, my new lab felt like it was a thousand miles away, and I missed seeing the notable faculty in the corridors. Not that I ever spoke to them, but I felt as though I could if I wanted to. In fact, one day a colleague introduced me to a famous professor when I needed to get some electron microscopy done and had no way to do it. The famous professor agreed to do it for me and even took the time to acquaint me with the workings of the electron microscope.

A few years later, after I had left that institution, the famous professor had become the president of the scientific society that I was a member of, and I bumped into him at the annual meeting. The previous day I had been bemoaning my feeling of isolation in my new institution to a friend, and he suggested I join a committee. He reminded me that the famous professor was president of the society and perhaps he could help me do that. So, when I bumped into the famous professor, I asked him whether he had any suggestions about how I could get more involved with the society. He immediately suggested that I join the finance committee, which had vacancies, and said he could just put me on it. I wasn't thrilled about this committee and had been thinking of something more scientific. Sensing my reluctance, he told me that finance was the key to everything that happened in the society and that it would put me in the middle of all that. I reluctantly agreed. During my ten plus years on the committee (I eventually became treasurer and a member of the executive committee), I met nearly every important scientist in my field by virtue of attending finance committee meetings, society business meetings, and meetings of the governing council. Over the years these connections were of immense help to me professionally.

Joining a committee in a scientific society is another path to getting your scientific and extrascientific skills recognized. This is true at any stage of your career but will be most valuable in your early years. Many scientific societies have committees specifically for early career scientists that are easy to join. As in institutional committees, these are great places to meet people outside of your existing network, including scientists who are prominent in your field who you may otherwise never get to meet.

STRATEGIC USE OF THE WEB AND SOCIAL MEDIA

Use these but with caution. Having a profile on LinkedIn is fine and is a kind of networking but only in the sense that standing still at a networking reception and hoping that someone will come up and talk to you is networking. LinkedIn is excellent for posting periodic updates about professional items of interest to your community. Recruiters and hiring managers will almost always look through LinkedIn for leads or to get more information on someone they just met or who contacted them. So be careful what you post. Do not count on LinkedIn as a primary source of job leads, but it does not take a lot of effort to maintain a profile.

Social media sites such as X and BlueSky can also be very useful for posting updates on your publications, technical accomplishments, conference presentations, etc. If you post about social or political matters, keep in mind that future hiring managers, recruiters, and members of search committees will as a matter of course search and examine your feeds.

Social media can also be useful if you want to establish an interest or credibility in an area that may not be obvious or even absent from your résumé. For example, if your résumé is principally filled with scientific accomplishments, but you are looking at jobs in academic teaching institutions, posting about and commenting on trends in science education is a way of showing that you are committed to and think about such matters. If your goal is to get a job in industry, posting about exciting trends or interesting companies can show that you are serious about the private sector and that you are following the latest trends. Doing so does not take a lot of effort, and in this day and age you would be remiss not to do so.

MAINTAINING YOUR NETWORK

Once you start building a network of contacts, it needs care and nurturing. Meeting someone once is not enough to make them a useful contact. You need planned maintenance to ensure that. Network maintenance is critical for both early career scientists as well as those who already have a job but think they may be moving on at some point in the future, even years from now. So that is basically everybody.

Maintenance means communicating periodically with your current contacts so that they remember you when you need them and so that you are in their mind when they hear about an opportunity that is a fit for you, even if you are not actively looking. Maintenance need not be a huge time commitment, but you do need to be conscientious about doing it. All that is needed is to keep in touch.

- A yearly email that says "Hi" with the most casual update on what you are up to
- A short email or call regarding something that they are up to or have done—congratulating them or recognizing a recent publication of theirs, for example
- An email about some news, publication, opportunity, or scientific result that might be of interest to them

The last two can be especially effective as they convey that you are interested in *them* and what *they are* doing, not just reaching out to them out of self-interest. If you start doing these things early in your career and keep it up, you will find over time that you have created an extended professional network.

Does this mean that when the need arises you should not reach out to people from the past who you have lost touch with? Definitely not. But if you have done this kind of maintenance, your chances of getting a response are better.

When I was a postdoc, I was working on a project that required human lymphocytes, which I did not have ready access to. There was a new assistant professor down the hall who worked on lymphocytes, and my adviser suggested I go talk to her, which I did. Over the next few months, she supplied me with lymphocytes, and we became friends. As the years passed, we both left that institution. I got an academic job, and she got a job in a startup biotech company in the area. Although we hadn't been in contact for a few years, one day I emailed her and asked if she would like to have lunch and catch up. I had no specific motive or objective but just thought it might be fun. At this time in my career, I had never given a thought to going into industry. As it turned out, going to lunch with Barbara became an annual event and it lasted for about ten years. At around year ten I decided that I wanted to make a major career change and move into the world of biotech. During my annual lunch with Barbara, I told her of my decision and asked that if she heard of any relevant openings to pass them along. No more than a week later I got a call from a recruiter who told me that he got my name from Barbara and that he was looking for a Director of Research for a local company. Long story short, within eight weeks I had a signed job offer in hand.

Commentary: This may not sound like a networking story, but it is. This person maintained his connection with Barbara for no purpose other than to maintain the connection. He did not have an extensive network, but he valued and kept in contact with the network he *did* have. Occasionally that pays off and is good enough.

HOW TO DO IT: NETWORKING SCRIPTS

Networking is not about getting a job, although that can be an outcome. It is about establishing connections with people so that when the *time comes* to get a job, they may be of use to you, or you to them. Savvy networkers know that the most useful connections are those that are mutually beneficial.

People you encounter at a meeting, a social event, or an actual networking event

With these people, you have no objective beyond meeting them, finding out who they are, and telling them a bit about yourself. And I do mean a bit. You need to have prepared the well-known elevator pitch that conveys the essentials about yourself in about a minute or minute and a half.

- Who you are: name, current position, maybe who you work for
- What you are working on: one big thing you are working on and why it's exciting

As you are doing this, you are watching their facial expressions and body language to see if anything you just said sparks interest or recognition.

With your introduction over, you now need to get into connection-making mode. The best and fastest way to get there is to get them to talk about their favorite subject—themselves. Ask questions about where they work, what they do, and how they got into their science. While they are telling you this, you are making eye contact, nodding, and saying things like "Wow," "Cool," and "Interesting." Then ask them about themselves even more. Ask them follow-up questions about what they just told you, get fascinated by the details of their science, and show immense interest in everything they say.

Based on what they tell you, try to come up with something you can share that might be of interest to them, either personally or professionally. It could be something you saw in a poster session, a recent journal article, or even a rumor. You want them to be helpful to you, but you need to *start* by being helpful to them.

Now it is time to pray that they reciprocate. Some will simply bask in the lavish attention you have paid them and wander off following the person with the hors d'oeuvres tray. Others will engage with you, and if you have enough in common, one of you will ask the other for their contact information, and you have just made yourself a contact. Pat yourself on the back.

Here is a case study involving someone we'll get to know later in this book named Samantha Jiminez, an analytical chemist. You can see her résumé in Appendix 1 of Chapter 3.

Samantha Jiminez is a Ph.D. analytical chemist who was laid off from her job three weeks ago as an analytical scientist at Pfizer Pharmaceuticals' Cambridge, MA, facility with no advance notice. She's been applying for jobs online, but it's still early and she has yet to receive any follow-up for any of the positions she's applied for. With time on her hands, Samantha decided that the meeting of the American Society for Analytical Chemistry, being held at the Boston

Convention Center, would be good to attend to try to supplement her job search. She has never attended an event like this other than as an employee of a company and is nervous about feeling like an outsider, especially having just been laid off. There is a part of her that feels like she is a failure.

On the second day of the meeting, she decides to attend a 4:00–6:00 p.m. networking event sponsored by a big pharmaceutical company. The event is being held in the ballroom of an adjacent hotel, and as Samantha walks into the huge room, a server walks up to her with a tray of wine glasses and offers her one, which she accepts. As she stands by the door sipping wine, groups of people wander around her. She does not know anyone here and feels embarrassed that her conference name tag just has her name on it, with no organizational affiliation. Everyone else's tags have names of recognizable companies or institutions on them. She feels out of place and considers leaving.

Looking around the room, to her left, she sees a table arrayed with a beautiful display of sushi and huge plump shrimp and her hunger gets the best of her. She walks over, takes a plate, and puts several pieces of sushi and shrimp on it and places them on a nearby stand-up cocktail table. Just as she puts a huge shrimp into her mouth and starts to chew, another attendee with a name tag that says "Josh Carter, Monel Institute, Philadelphia" walks over to the table, places his plate on it, and says, "Mind if I join you? These shrimp look terrific." Samantha's mouth is full of shrimp, so all she can do is smile and wave her hand indicating it's OK. Josh laughs and puts a shrimp in his mouth and mimics her hand gesture and they both laugh.

After they finish chewing, Josh extends his hand and introduces himself. They shake, and Samantha says, "I'm Samantha, great to meet you."

Josh looks at her name tag and says, "You must be here in stealth mode—your company's so secret they won't even let you put their name on your name tag."

Samantha smiles, makes a thumbs-up gesture, and says, "Wow—that is a great idea—I've been wondering what to say about that name tag all day and you just handed me my line!"

Josh raises his eyebrows and says, "You mean that's true?"

Samantha responds, "If only. Actually, if I were at this meeting a month ago, my tag would have said Pfizer. I was there for five years and last month they closed their Boston analytical group because they are shifting a lot of that work to their Ireland facility. So, here I am, your basic freelance analytical chemist."

Josh looks interested. "Gee, that's too bad, and I'm sorry to hear that. So, what's your specialty? What did you do at Pfizer?"

Samantha replies, "Well, I mostly worked on proteins, peptides, and conjugates. What was really nice about that job was that I was in charge of putting in place their new mass-spec characterization protocols for protein conjugates and also was part of a great team that built their high throughput analysis platform for small peptides. We were actually moving to do the same for small molecules next spring and had even bought all the equipment. Now, all of that is being transferred to the Ireland facility. Truthfully, I feel pretty good about what I did there, even though others will be taking it over. I'm not complaining—it was a great place to work and the people there were absolutely first rate. I guess I'm ready for the next big thing in my career."

Josh has a thoughtful look on his face. "Well, how do you feel about Philadelphia?"

In this story, Samantha conveyed a lot of important information about herself in the course of a short, casual conversation. At no point does it feel like Samantha is desperately hunting for a job or seeking to "collect" contacts. In the course of no more than a minute or two, she managed to convey several important messages to Josh:

- Who she is
- Her technical specialty
- A few key accomplishments
- Looking for the next step in her career

Just as important, however, she also conveyed that she is a team player ("part of a great team"), enjoyed her work at Pfizer, and bears no ill-will toward her employer. Rather than being defensive about not having Pfizer's name on her badge, Samantha's honesty about her discomfort helped her establish a personal connection with Josh. In short, she sounds like a very positive person. It is no wonder that Josh's response is "How do you feel about Philadelphia?"

The next step for Samantha is to reciprocate Josh's interest. "I hear Philly's a great place to live. What do you do at Monell?"

People who you think may have a job available right now or in the future in their group or organization

How *not* to approach: "Hi, Dr. X. I wonder if you have any openings in your company/lab/group right now?"

This violates the first rule of networking—it is not about getting a job, it is about making connections. Ninety-nine percent of the people you will talk to will not have a job available at that time. So, when you ask them if they do and they say no, it is going to feel like the conversation is over, which is the last thing you want.

Better approach: "Hi, Dr. X. a) I was just at your company's booth in the convention hall, b) I just saw a poster from your group, or c) I just read an article in *Cell* from your team, and a) I had a question I wonder if you could answer, b) I just wanted to tell you how cool that result is—it's something I've been interested in for quite a while, or c) I wanted to let you know about a related poster I just saw today that I suspect you are already familiar with from Dr. Z's group in Toronto. It's about ..."

Now you have their attention because you are talking about them, their work, or their company. They will be interested in talking to you to the degree that they think there is something interesting or useful in it for them. You need to have a plan to accomplish that—know something about them, their group, or their company, even if you just learned it two minutes ago at a talk or a poster session. Then you need to let them know that you find them, their work, or their company fascinating in the extreme. This will get you five minutes of airtime with them, during which, if you are lucky and you have not dropped a piece of sushi on their shoe, they will ask you the Golden Question:

Tell me a little about yourself.

Of course, it may not come out exactly like that. It may sound more like "Oh, do you work in this field?" or "So, what brings you to this meeting? Are you presenting something?" or "It sounds like you know this field; where do you work?" This is your opportunity to "tell them a little about yourself" in about a minute and a half. This is your "elevator pitch" that you have practiced till you are sick of hearing yourself talk, yet—and here's the hard part—every time you give it you need to sound like you are just making it up on the spot for this specific person.

So, what's your story? Here are the things you need to cover in less than ninety seconds:

- Who you are: name, current position, maybe who you work for
- What you are working on: one big thing you are working on and *why it is exciting*
- Something about your situation: "I'm just finishing my postdoc and getting ready to go into the job market." Note that this is very different from "I'm looking for a job," which may cause them to shut down. "Getting ready to go into the job market" is not a question for them but rather a statement about you. Another is "I'm thinking about looking into moving into a job in industry." Again, no asks and no pressure, you are just talking about yourself.

With any luck at all, they will make a noncommittal comment like "Cool" or "OK, that sounds interesting." Then you go to phase two, which is the real reason you are talking to them. "Yeah, it's pretty cool. I'm hoping to finish up in about six or nine months and start looking for a job." Here are some possible responses:

1. They suddenly look at their watch and say they are late for a meeting. This is your next opening. "It was really great to meet you. If it's not too much to ask, at some point I would really appreciate the opportunity to speak with you again. I'm just trying to map out career options and next steps, and I think the more people like you I talk with the better prepared I'll be when I start looking, so it would just be a very general conversation." They may demure, but they may also say "OK." Congratulations, you have made a contact.
2. They ask you some follow-up questions about what you might be looking for and then tell you that they or their company are not hiring anytime soon but they wish you good luck as they look at their watch and tell you they are late for another meeting. Again, this is your opening. You say, "Oh, of course. I understand. Is there any chance there is someone you know here or who I could email who might be willing to talk to me about what it is like to do this kind of work at your company/university/institute? I'm just trying to network and to educate myself." They may say something like "Not off the top of my head." Which is actually terrific because now you can say "Oh, I understand perfectly. Would it be OK to send you an email in a month or two just to check in if you might be able think of someone I could talk to?" Maybe you are being a pain in the neck, but that is your new job.
3. They say, "There may be something coming up in my department. Tell me a bit more about yourself." OK, this almost never happens. But on the 1% chance that it does, be prepared to deliver your one-minute elevator pitch follow-up.
   - What your major scientific and technical skills are (very briefly)
   - What you think you would like your next career step to be (i.e., something that looks very much like their company, lab, etc.)

What happens next of course depends on their response. If they ask you follow-up questions, then you are on a roll and you follow their lead. If, on the other hand, their eyes glaze over and they start scanning the room for anyone else to talk to but you, you pivot back to asking them about their favorite topic—them. Show them you are as interested in them as you wish they were in you. "So, what's your story? How long have you been with Gorilla Pharmaceuticals?" or "What's it like to work at Famous University in your department? I've heard a lot about it but have never visited there," or "How long have you been in industry? I've been thinking about looking for my next job there. What advice would you give me?" If these ploys fail, you have given it your best shot and it is time to cut them loose and move on to someone else.

If they seem open to hearing about your deepest passions, interests, and career plans, this could mean that they find you truly fascinating or that standing there and listening to you is just easier than moving around and trying to find someone else to talk to. In either case, you need to seize the moment. Here are the talking points you need to be ready to cover during the ensuing conversation.

1. You want nothing from them. They will find an excuse to walk away if they feel pressured that you are going to ask them to actually do something for you.
   - "Right now, I'm just networking. Looking at options, talking to people."
   - "I'd love to hear about your experiences at company/university X," "Did you do a postdoc, or did you go right into industry?," "Do you think that's still the right way to go in this climate?," or "I'm trying to decide if I should do a postdoc or just go on the job market." (Then close your mouth and listen.)
2. Who you are. You already covered some of this in your intro, so add one or both of these:
   - Your strengths. "I've been told (much better than "I am" or "I have") that I have pretty good analytical/lab/coding skills. I have a poster up tomorrow if you get a chance to drop by."
   - A couple of achievements you are proudest of. "I'm actually pretty excited about creating this cool algorithm that filters and analyzes neural signaling to create three-dimensional and temporal maps."
3. Be relentlessly positive.
   - Do not be negative about anything. Not your current job, department, or grants. Nothing. You love what you do. But as is only natural, it is time to move on.
   - Signal that you are open-minded about exactly how you can contribute wherever you go. "I'm pretty entrepreneurial and not afraid of working hard, learning, and adapting."
4. Engage them.
   - "What exactly do you do?"
   - "What do you like about your job/company/department?"

- "How did you get your position? How long have you been there? Where were you before that?"
- "Do you think you made the right choices when you were at my stage? Would you have done anything differently?"

5. Do not go here:
   - Your ideal job, unless they actually ask, "What's your ideal job?"
   - Do not get into your geographic preferences, desired salary, and other specifics. If they ask you "Are you willing to relocate?" or "What kind of salary are you looking for?" say "I'm really open to anything right now and I'm pretty flexible as to where I live." (That may not be entirely true, but adding caveats may blunt their interest.)
6. Leave them with:
   - You are upbeat, enthusiastic, energetic, and fun to work with.
7. Follow up. Help them keep you in mind.
   - Send them an email thanking them for chatting with you.
   - If someone gives you a name and you follow up with that name, let them know you did it and thank them.
   - Make a spreadsheet of people you have spoken with or met. Update them on your progress periodically (once every three months if they are helping you with an urgent job search or once a year otherwise).
8. Big truths about networking:
   - Your most valuable contacts are people you already know or who know you. While they will give you your best leads, and your best references, weak or casual contacts are also important. Nurture both.
   - Existing contacts will lead to others, who will come in unexpected ways that you cannot predict. Do not filter too much who you talk to.
   - The measure of a successful meeting is that you got at least one new contact. "Can you think of one or two people who I could talk to about their experiences? I'm trying to learn as much as I can while I'm here."
   - Treat this process like a research project. It will take time and requires many small steps, each of which will contribute to your overall progress.
   - You are selling your work and you are also selling you. Most of us forget the "you" part when we are chatting people up. Whether you are thinking of academia or industry, people hire people not research projects.
   - Smile. Not too much but just enough. Even while you are on the phone and they cannot see you and especially on Zoom because they can see you.
   - Find something to like and admire about everyone you talk to, even though you do not know them very well. Comment on those things.

USING YOUR NETWORK TO GET A JOB: FOCUSED NETWORKING

If it is crunch time and you are actively job hunting, you are into the realm of focused networking. You have a specific objective, and you are working against the clock. This is where all the network maintenance you have been doing pays off because now rather than just "checking in," you are letting them know that you are in the market for opportunities. Now you need to treat networking like a project *with a deadline.*

Your networking project plan

- Make a list of everyone you know and put them in categories:
  - People you have never spoken to or met, but you think you could introduce yourself to them and they might know others to introduce to you
  - Anyone you have ever talked to, no matter how casually
  - Friends and family
  - Peers and colleagues (past and present)
  - Mentors and advisers (past and present)
  - Current and previous employers
- Develop a timeline (see the one at the beginning of Chapter 3 for an example). Figure out how many people in your network you should get in touch with each week during your search to cover all or most of your contacts. If you have eighty contacts and twelve weeks allotted to Phase 1, that is about seven people a week.
- On your spreadsheet, keep track of who you have contacted, when and how you contacted them, and whether they responded. Also include when you should contact them again.

What does "get in touch" mean?

It means sending them an email, having an in-person conversation (even a casual chat in the hallway counts), buying them a latte, going to lunch, or setting up a phone or Zoom call. Now it is OK to let them know you are in the job market and you should do so.

This is the simplest part of networking. All that is required is something like this:

Hi, Dr. Famous:

It was great to meet you at the IEEE meeting last month. I'm just writing to let you know that I'm seeking a position as a (position you are seeking) starting sometime in (approximate date). If you hear of anything that might be suitable, I'd appreciate it if you could pass it along. Also, if you know of anyone who I might reach out to who might know of something, I'd be most grateful if you could share their contact information. If you think it would be helpful, I would be happy to send you my résumé (or CV). Thanks so much for any help and any introductions you can provide.

Sincerely,
You

When I coach scientists who are in this mode of networking, they often ask how much time they should be spending on it. My answer is that if they do not feel that they are spending too much time networking, then they are not spending enough time. Because networking is hard and takes many scientists out of their comfort zone, even a small amount feels like work. Do not let feeling burdened or uncomfortable deter you. Networking is a project, not a hobby, and if you do not set goals, milestones, and timelines, you will never get to the ultimate deliverable—a job.

CHAPTER 2 SUMMARY

Importance of networking

- While a fortunate few may have opportunities come easily, most scientists need to actively build and maintain professional networks.
- Networking is crucial for accessing job opportunities, advice, and career advancement.
- It serves as the primary pathway through which most scientists actually find jobs and advance their careers.

Systematic approach to networking

- Networking must be treated as a project requiring systematic effort.
- Key activities include attending conferences, joining committees, and maintaining regular contact with connections.
- Success requires practice and consistency rather than hoping work will speak for itself.

Practical networking strategies

- Detailed scripts and strategies for different networking scenarios are presented.
- Specific applications include casual introductions at conferences and focused outreach when actively job hunting.
- A key takeaway is to be genuinely interested in others rather than appearing desperate for opportunities.

Building vs. maintaining networks

- *Building new connections is important at the outset*: through scientific meetings, poster sessions, and professional organizations.
- *Maintaining existing connections is important on an ongoing basis*: through regular communication, periodic emails, updates, and mutual support.

Key principles

- Networking is an essential skill that requires practice and consistency.
- Focus on genuine interest in others and mutual benefit.
- Finding positions through your network is more effective than relying solely on online applications or assuming technical excellence alone will lead to opportunities.

3 It's Not About You

Cracking the Code of Academic and Industry Job Hunting

If you have completed graduate school or a postdoc and are now ready to look for your next position, congratulations. You probably feel a mix of accomplishment and relief plus a bit of anxiety about finding a postdoc, faculty, or research scientist position. Pretty much everyone else in your position feels that way too. It is natural that you will be nervous about your job interview and the seminar you may have to give. These are stressful situations, and you probably have little prior experience in them.

The following chapters will get you prepared.

PLAN BACKWARD, START NOW: YOUR JOB HUNT TIMELINE

Your job search requires a plan, and all plans start with a timeline. How long is this going to take? When should I start? The answers to these questions are: Probably longer than you think, and you should start now. As with networking, treat your job search as a project.

Start by figuring out when you need to finish then work backward. If you need to be working in eighteen months, how long will it take to do all the things that will get you a job in that time frame? Once you know that, you can decide when to start.

If you are an amazing candidate in a strong job market, it can take as little as three or four months. So, if you need to have a job in eighteen months, you do not need to start until over a year from now. If you are merely a strong candidate, have a good publication record, are a great speaker, and a few people have expressed interest in you as a candidate, starting six to eight months in advance might work. If you are neither of these candidates, the table below suggests that it could take anywhere from eight to

Typical total time to get a job offer

Amazing candidate in strong job market: 3–4 months
To be working in 18 months: Start 14 months from today
Average candidate in average job market: 8–14 months
To be working in 18 months: Start 4–10 months from today
To be working in 9 months: You should already have started

fourteen months. So, to be working in eighteen months, you need to start between four and ten months from today.

There are other considerations. If you are applying for a job in industry and get an offer, unless it is for a senior management role, they are going to want you to start within weeks. You are not going to be able to ask them to wait six months so you can finish your current projects. So, if you do need to finish current projects, you need to factor that into your timeline.

You may have more flexibility in an academic institution, but do not count on it. If you are hired by a teaching institution and they make you an offer in June, they may be expecting that you will start teaching in August, and you may not be able to negotiate your way around that.

Job search timeline for average candidate in average job market

Cumulative months		Steps in the process	What you should be doing
	Today and earlier	**Prepare**	• Reach out to your network and scan job postings. • Draft your CV for academic jobs and résumé for industry jobs. • Draft your application letter. For academic jobs, draft research description and teaching statements.
3–6	Phase 1 3–6 months	**Search**	• Customize your résumé and CV for each job you will apply for. Same for research and teaching statements for academic jobs. Have them reviewed by trusted advisers.

(Continued)

Cumulative months		Steps in the process	What you should be doing
6–12	Phase 2 3–6 months	**Send applications.** If people have said that you are a stellar candidate, you are lucky, and the job market is strong, this could be a 1–3-month process. Otherwise figure 3–6 months.	• Do extensive research on candidate institutions. • Prepare, customize, and practice (many times) your job talks for academic jobs. • Review and practice answers to interview questions. • Prepare draft startup budget for academic jobs.
10–16	Phase 3 4 months	**Interview.** We are using 4 months as a placeholder between when you start interviewing and when you get your first offer. That is optimistic and will include one or more on-site visits.	• Finalize your startup budget. • Review negotiation tools. • Thoroughly research the people you will be meeting with.
10–17	Phase 4 0–1 month	**Negotiate job offer.** For high-level jobs this can take a month or more. For low-level jobs there may be little or no negotiation.	• Review negotiation tools again. • Review offers with your trusted advisers.
10–18	Phase 5 0–1 month	**Accept job offer.** For high-level jobs you can take a week or more once an offer is made. For others they may want your decision in days.	• Make sure you really want this job before you say yes.
11–19	Phase 6 1 month	**Processing, paperwork, and onboarding.** They will take longer than you might expect.	
12–20	Phase 7 1 month	**Start the job.** The lower the job level the sooner they will want you to start. For more senior positions it is not unusual to be able to put off the starting date for many months.	

This is where having a good mentor will pay off because they will give you honest guidance about your job prospects, the appropriateness of the positions you are thinking of applying for, and the current job market. This will help you create your own timeline.

The table above is a typical timeline for an average candidate, but it is just an example. Your timeline may be different. You could get lucky, and the process will happen a lot quicker than you expect, but planning to get lucky is not a good strategy. Better to plan to be unlucky.

IS THIS JOB FOR YOU?

The first question to ask yourself before applying for a job is "Am I qualified for it?" Once you have read a job posting and the candidate qualifications or requirements, ask yourself if this is true. Be honest with yourself when making this evaluation, something that some of us are incapable of. Show the job description to somebody you know—a friend, an adviser, or a colleague—and ask them if they think you are qualified. If you get a lot of uncomfortable squirming and obfuscation, you have your answer. Do not apply. If you really do not have most of the qualifications, background, and experience needed, no amount of résumé optimization is going to help you get it. This is not to say that you should not be aspirational, but you should be realistic.

You may think "These résumés are going to get screened by some AI bot so what's wrong with taking a chance? Maybe there's something in my résumé that even I don't see that this company or organization will value." If indeed the résumés are going to be scanned by an AI bot (which in many cases, they will be), then by all means send it in.[24] Send as many as you like. Ask your friends to send their résumés in the hope that you will crash their server and teach them a lesson for scanning résumés with AI bots.[25] If on the other hand your résumé is going to get reviewed by an actual person in a university department or small startup company, do not burden them by sending them something that is going to end up being noise.

It is just as important not to make the opposite mistake—concluding that you are not qualified if you do not match 100% of the position requirements. As you will learn, you do not need to be a *perfect* match with who they are looking for, only a good match. Women may be less likely than men to apply for jobs that they think they are not a perfect match for.[26,27,28] In one study, almost 22% of women chose not to apply for a job that they thought they were not 100% qualified for compared to only 13% of men.[26] So, before you decide *not* to apply, do as I suggest—show the job description to a friend, tell them why you have decided not to apply, and see if they push back.

The second question to ask yourself is "Do I really want this job?" Is it the right job for you? Is it the right place for you? Will you be happy there? Is there even a chance that you and/or your partner would consider moving to the place where this job is? If you cannot answer yes to these questions, think twice about sending your résumé in.

Some people will advise you to apply for jobs that you really *do not* want as practice for applying for jobs that you really *do* want. There are going to be actual people spending time reading your résumé and talking to you on the phone. If you do not really want this job, you are wasting their time. We are now in the realm of ethics. You want to do it because it is good practice for you, but it may be bad and inconsiderate for the people you are sending your résumé to. Do such a thing once or twice but no more. If you feel like you are still not ready for a high-stakes interview, practice with friends, lab mates, or colleagues.

YOUR EMPLOYER HAS NEEDS

If you are contemplating a job in academia, you hopefully already know that positions even in mid-tier academic science departments are incredibly competitive with dozens, or more likely hundreds, of applicants. Industry is no better. If you have had a charmed graduate school or postdoc experience with impressive publications in fancy journals and you have trained in a world-renowned lab under a helpful mentor who has introduced you to their friends who are interested in your candidacy, congratulations, you are off to a great start. Yet I have seen candidates like that who after six months on the job market with numerous "nibbles" (phone calls or screening interviews) still have no offers. They are wondering what they are doing wrong. In many cases, the answer is that they failed to recognize the one immutable rule of job hunting:

It's not about you, it's about them.

Put yourself in the position of someone on the search committee for the position you are applying for. How would you go about choosing the best candidate? Only when you reverse engineer your job search by engaging in this exercise will you understand what we mean by "It's not about you, it's about them."

As a hiring manager or member of a search committee, your overriding interest is in hiring someone who meets the needs that you, your lab, your department, or your company has. So, the interview process is an exercise in matching the candidate's skills, capabilities, and experience against your needs. Now go back and be yourself—the candidate. With this new insight into what the search committee is up to, do you now know what you need to do to make yourself attractive? How about "I need to carefully review the position description and, if possible, find someone inside the organization I can talk to, to make sure I have an accurate understanding of what they are looking for. Then, I need to customize my cover letter, résumé, and generic research seminar so that they highlight exactly how I meet *most or all of their needs*. I need to focus on what I can do for them, rather than on what they can do for me"? If this is your answer, pass "Go" and collect $200.

Note that we did not say "meets every single one of their needs." The position requirements in a job posting is a wish list. While the organization may wish to find a candidate who matches the list perfectly, in nearly all cases no such candidate exists. And despite what you may think, as you read the very specific list of candidate requirements, the hiring organization knows this. Also note that even if you think you *do* meet all or most of their needs, you need to prove it to them.

Hiring managers or search committees will not see your inherent value simply by reading your résumé and cover letter. You need to spell it out for them and help them put the pieces together. The people reading your application may be reading twenty-five, fifty, or even one hundred others in a short time frame. Make their job easier by being clear, specific, and concrete.

Many candidates who actually have most or all of the requisite skills do not make the cut because they have failed to demonstrate this in their résumé, during their interview, or in their seminar. They fail because their focus is solely on showcasing *their* accomplishments and not on highlighting how they can meet the *employer's* needs.

This misdirected focus is understandable in those emerging from academia where success is predicated almost entirely on individual accomplishments. By the time they are ready to look for a job, most academic trainees can talk about what they are working on and what they accomplished. But few have any experience in spelling out how their skills and capabilities match the needs of an employer.

The way for them to spell it out is to customize their résumé or CV, their cover letter, and their research and teaching statements to show how they meet the employer's needs. And to do that they need to understand the hidden and implicit meanings of the words and phrases used in the position description.

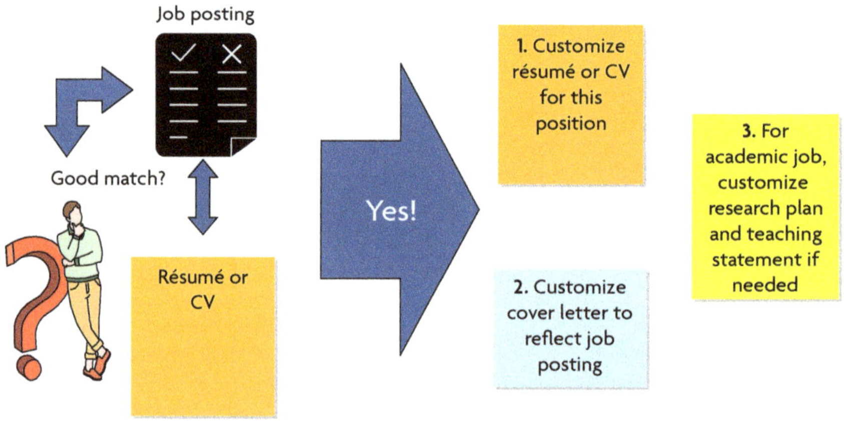

GET IN THEIR HEAD: THE HIDDEN MEANING IN INDUSTRY AND ACADEMIC POSITION DESCRIPTIONS

The aspirational leading paragraph

We have all seen position postings that start like this: "Seeking highly motivated team player to contribute to cutting-edge research team developing world-class disruptive technologies to solve humanity's most pressing needs. Key personal skills needed include: exceptional work ethic, ability to think critically and act strategically, excellent written and verbal communication skills, exceptional time management and multitasking skills. The ideal candidate will have documented expertise in the scientific areas specified evidenced by first author publications in high impact journals."

Does this mean you need to start your letter or résumé with "I am a highly motivated team player who seeks to contribute to a cutting-edge research team developing world-class disruptive technologies"? This could prompt their screening program to flag your application as "possible," but once it gets read by a human it is going to sound silly. And unless your application is truly outstanding, it may cause them to put it aside.

You need to convey these things in your own words and in a way that is different from the position description, e.g., "My employers consider me to be a critical and strategic thinker with an exceptional work ethic. I am attracted to this position because it will allow me to join an innovative research team, work with groundbreaking technologies, and help improve human health."

You will need to convey these same characteristics when you meet prospective employers and then have examples, stories, or anecdotes that illustrate these traits. This includes things you have done or ways you have demonstrated motivation, critical thinking, and time management. That will get their attention far more convincingly than having your résumé restate verbatim what was in the position description. You will see how to do that in Chapter 6.

Detailed or specific candidate requirements

Nearly all position postings have a list of requirements or qualifications for potential candidates. These are usually preceded with qualifiers like "must have" or "preference given to candidates with." The table below decodes the embedded meanings in the most common phrases in candidate requirements.

Deciphering candidate requirements

Phrase	Meaning	Strength (10 is highest)
"Must have minimum of 3 years of experience in . . ."	Not only do they want you to be experienced in this, they also want a specified number of years of experience. This looks like it is a "must have" skill. But you can assume that the specified years of experience is going to be a soft number. Will they really not consider a candidate who has only 2 years of experience if it was intense hands-on experience and they can prove they know the material? I doubt it. So, this gets a 10 for strength of the requirement but a 5 for the number of years part. What they are really saying is that you need to know this. The 3 years of experience part is the easiest way for them to determine that. You may be able to convince them that you have the experience in some other way.	10/5
"Essential requirement" or "Essential skill"	You really need this skill. They are signaling that if you do not, you should not apply. However, if you have other impressive or compensatory skills, and you want to be bold, go ahead and apply. This is the only case where you can mention up front that you do not meet this requirement since you are going to explain to them why they should hire you despite that. If you are convincing, this could get their attention.	9

(Continued)

Phrase	Meaning	Strength (10 is highest)
Anything listed without a modifier in the "Candidate requirements" section	The absence of a modifier like "essential" or "must have" signals that while this is important, it is not a deal breaker. No need to mention that you do not have this in your letter.	7
"Prefer candidates with . . ."	You really do not need this, and it is not a requirement, but candidates who have it may receive preferential consideration. Not having this is no reason not to apply, and you should not mention it.	5
"Skill X, a plus . . ."	You do not really need skill X to do this job, but if you have it, it could bump you up in the queue.	2

What if a skill you do not have is listed as an *essential* requirement but you believe that you can acquire it? In that case, do not bring it up. If *they* bring it up, you must be honest. But do not confuse honesty with transparency. Honesty means that *everything you say is true*. Transparency means *you say everything that is true*. You definitely do not want to do the latter. Focus on your strengths, not on your deficits. Including phrases in your cover letter like "While I don't really have any formal teaching experience in …" or "While I have only two years of experience in …" are unnecessary.

The most useful phrase for such situations is "I'm a quick learner. I have complete confidence I can pick that up in a very short time." Even if the skill that you lack is one that is listed as an "essential prerequisite," you may be able to convince them that you have other outstanding qualities that compensate for the absence of the essential skill.

Deciphering an industry job posting

Below is an industry job posting for an analytical chemist copied and anonymized from an online service. Unlike a job posting for an academic position, which we will cover next, industry postings are relatively straightforward. Most consist of a) an aspirational statement of personal characteristics desired, b) the qualifications and skills the candidate should have, and c) what the candidate needs to be able to do.

In the following, the elements of the posting are on the left and what each element means is on the right. As discussed earlier, even qualifications that are listed as essential may be subject to negotiation.

Deciphering an industry job posting

Hyperbole Pharmaceuticals seeks an analytical chemist	*What it means*
Position Description • Hyperbole is seeking a highly motivated analytical scientist to join its Advanced Drug Delivery team in Attleboro, Massachusetts.	*This is a straightforward description of the position but as with most such postings, it is hard to know what this job looks like on a day-to-day basis.*

(Continued)

Hyperbole Pharmaceuticals seeks an analytical chemist	*What it means*
• The candidate will be responsible for the analytical method development and characterization of peptides, proteins, small molecules, oligonucleotides, and advanced drug delivery platforms. • This position will support therapeutic programs in multiple disease areas and will contribute to the development of novel drug delivery. • The position involves the handling of chemical compounds and the preparation of small molecule and large molecule samples using manual and liquid handling robotics, use of HPLC, HPSEC, various spectrophotometric and other analytical methods, documenting and reporting results in electronic notebooks. • This position reports to the Head of Advanced Drug Delivery (Drug Product Research). Your results will be captured in an electronic lab notebook and communicated to other team members to support our projects. Our labs are a fast-paced and exciting environment and your data will be key to decision making.	
Essential Functions • Utilize subject matter knowledge in analytical chemistry to develop and verify analytical methods to quantify and characterize peptides, proteins, small molecules, and drug substances in a variety of formulations. • Develop analytical technologies and methods to increase current analytical capabilities for therapeutic molecule testing including advanced separation and detection methodologies. • Generate product supporting data packages by applying HPLC, UPLC, LCMS, GCMS, SEC-MALs, MS and other instrumental analysis techniques. • Collaborate with in-house teams to define and validate dosage and chemical attributes of the drug candidates. • Operate, maintain, and troubleshoot analytical lab instruments including HPLCs, UPLCs, and MSs.	*As above, there is nothing subtle or hidden here. Just a factual though vague description of what the position entails.*

(Continued)

Hyperbole Pharmaceuticals seeks an analytical chemist	*What it means*
Qualifications Bachelor's degree required. Degree within subject matter expertise (biological, biochemical, or pharmacological sciences) preferred.	*This is pretty flexible, other than a bachelor's degree being a requirement. The scientific disciplines are broad enough to cover many areas, including possibly chemistry.*
4+ years' relevant technical experience required or master's degree within subject matter expertise can be considered.	*Since there are no qualifiers (i.e., "a must"), assume that the "4+ years" is flexible. If they are saying that a master's degree plus no relevant experience can substitute for 4+ years of experience, they would likely consider someone without a master's and only three years of experience, or even two if the experience was especially relevant to this job.*
Strong knowledge and expertise in analytical chemistry, including method development for biologics and advanced drug delivery technologies.	*"Strong" means different things to different people. Do not be put off by it.*
Demonstrated experience in evaluating and implementing new analytical technologies in drug analysis, formulation, stability.	*"Demonstrated experience" means that you did these things but leaves open to interpretation how often. "Implementing and evaluating" are fuzzy. They could mean you were shown some new technology; you gave your opinion of it (evaluating) and then you used it (implementing). Lots of wiggle room here.*
Proficient in data analysis and interpretation for drawing conclusions and making recommendations.	*Easy. If you have got any scientific experience you have to be good, if not proficient, at these things. Not a high bar.*
Experience in project management and working collaboratively with cross-functional teams.	*It is unclear whether "expertise in project management" means they want you to have run a project or simply that you are familiar with how projects are run and have been on project teams with members of other scientific disciplines. Start by assuming it is the latter, and if they really want you to have run a project that will come out in the interview.*
Excellent scientific communication skills and ability to engage with stakeholders effectively.	*This is a higher bar than it may seem. Most scientists do not have "excellent" communication skills and will have a hard time checking this box.*

(*Continued*)

Hyperbole Pharmaceuticals seeks an analytical chemist	*What it means*
Have strong attention to detail and ability to document experimental work electronically.	*If you cannot check this box, your scientific future is limited in general, not just for this job.*
General knowledge of cGMP, GLP, and regulatory requirements for pharmaceutical industries a plus.*	*The word "general" is the tip off. Anyone can gain general knowledge about almost anything after an hour on the internet. Also, it is a "plus" not a must have. If you have it, it may distinguish you from others. If you do not, it will not necessarily kill your prospects.*

*cGMP refers to current good manufacturing process, and GLP to good laboratory practice. These are standards set by the US Food and Drug Administration for the conduct of laboratory work relating to and the manufacture of therapeutics.

Deciphering an academic job posting

Below is a job posting for an academic Marine Genomics scientist based on one found online. The posting has been modified to make it for a position in the fictional Institute for Marine Genomics at Cornell University. As above, on the left are the elements of the posting and on the right are what each element means. Unlike the industry posting, which is relatively straightforward, this one contains a lot of nuances.

Academic position posting and analysis: Institute for Marine Genomics

Position description	*What it means*
The Institute for Marine Genomics (IMG) at Cornell University seeks a nontenure-track faculty member at the assistant professor rank. While the expected rank for this position will be assistant professor on the nontenure track, final rank and tenure status will be commensurate with selected candidate's experience.	*They would probably like to hire you for a nontenure-track position, but if you are a stellar candidate they will consider a tenure-track role. They are tipping their hand and letting you know you can negotiate the type of role it is. You will most definitely not mention that you would only consider a tenure track position in your letter. That time will come later when they signal that they might offer you a position.*
We are interested in candidates with a research program in marine genomics with a focus on marine ecosystems as well as organismal adaptation. Research focused on the genetic makeup of marine organisms, their environments, and the interactions between them is of particular interest.	*The message is that they are most interested in marine genomics and would prefer a focus on ecosystems and adaptation. Then they go on to list areas of "particular interest." Despite that list, this is actually a pretty wide window. If you fit any of the specific areas listed you should highlight that in your letter and your research statement. If you don't you won't say so. Rather you will describe your particular expertise in a way that is as relevant as possible to one or more of these areas.*

(Continued)

Position description	*What it means*
It is expected that the candidate's research program will complement ongoing projects within IMG and the university as a whole.	*In this and the following section they are signaling very clearly that they are especially interested in collaborative candidates. This statement does not show up in every job description, so you need to pay particular attention to it. Section five of your letter (see Chapter 4) is the place to mention areas where you see collaborative opportunities, including mentioning the names of possible collaborators in the department.*
The candidate will be expected to teach one or more of our core course offerings such as Marine Functional Genomics and Introduction to Marine Genomics.	*Since teaching will be part of the job, this signals that you are going to have to supply a teaching statement and possibly do a teaching demo.*
IMG has seventeen highly collaborative faculty members with research interests in the areas of marine genetic variation, conservation, marine microbiome, and evolution. IMG has a significant infrastructure to support research-based high-throughput sequencing and complex data analyses, including extensive bioinformatic and computational capabilities available to all faculty.	*More mention of collaboration as an important element.* *It could be useful to allude to how their infrastructure will be of immense utility to your research program.*
Qualifications	***What it means***
The candidate should hold a Ph.D., M.D., M.D./Ph.D., or equivalent in marine genomics, microbial genomics as applied to marine organisms, computational biology, computer sciences, genetics, or a related field.	*If you have the specified degree in the fields they list, great. The key words in this sentence are "or a related field." Clearly, particle physics is not a related field, but any number of others could be spun as one. Examples include marine biology with molecular/genomic focus, bioinformatics, molecular biology with marine applications, microbiology (general or environmental), evolutionary biology with genomic emphasis, oceanography with biological focus (biological oceanography), environmental genomics/metagenomics, systems biology, biostatistics, marine ecology with molecular components.*

(Continued)

Qualifications	*What it means*
Candidates should have relevant postdoctoral experience and a strong record of academic achievement as determined by publications and grant funding.	*As above, "relevant postdoctoral experience" is a pretty wide window. Describe your experience focusing on areas of intersection with their major interests. Similarly, "strong record" of academic achievement is in the eye of the beholder. A few extremely relevant publications or a larger number of moderately relevant ones could get you there. The last sentence about grant funding, which also shows up in the next qualification, clearly signals a preference, but thankfully for most of you it is not a requirement.*
Preference will be given to candidates whose research efforts are currently funded.	*Who wouldn't give preference to a funded candidate? How about no one? But note their wording—preference is not the same as "this is an absolute requirement." And "preference" means "all other things being equal," which they almost never are. If you are in all other respects a stellar candidate and meet all of the other requirements, you could edge out the funded candidate whose program is only marginally relevant and who does not signal that they are eager to collaborate.*
Instructions	***What it means***
Please submit, as a single PDF document, your curriculum vitae, brief statement of research interests and vision, a one-page teaching statement, and contact information for three references. Candidates should submit their application to . . .	*As with anything labeled "instructions," you must adhere to their wishes rigorously. Sending in three separate PDFs instead of the one they request is a sure tip-off that you are not paying attention, do not follow directions, or just do not care.*

Beyond what the position descriptions and candidate requirements state explicitly, there are a number of implicit requirements and expectations for both academic and industry jobs that may not show up in the job posting.

Hiring committees in *academia* are interested in your:

- Ability to get funded (this is most important for R1 institutions but relevant to others as well)
- Teaching and mentoring skills
- Being a good departmental citizen by serving on committees
- Having a research program that will be attractive to graduate and/or undergraduate students

Hiring committees in *industry* are interested in the following:

- Whether you are a team player, as evidenced by having worked as part of a team, and understand the difference between being an individual versus team contributor
- Whether you are familiar with the company or organization and committed to their goals and mission
- Your willingness to work on what the company needs rather than on what you are interested in

These are just partial lists. We will add a lot more in later chapters specific to each of these domains. Knowing what these implicit needs are, you can highlight them in your application materials and in response to questions during interviews.

If you have gone through job postings and found one or more for which you believe you are a good fit, it is time to start getting your application material together so you can get noticed.

RÉSUMÉS AND CURRICULUM VITAE: KNOW THE DIFFERENCE

To get noticed you need to make the case that your qualifications match their needs. If you do this convincingly, you will move to the next step—the screening interview.

You will need to prepare and send different documents for industry and academia. These documents need to be customized for every job you apply for and are listed in the table below. We will cover each of these documents as well as how the one common document, the cover letter, differs between industry and academia.

Documents needed for your job application

Industry	Academia
Résumé	Curriculum vitae (CV)
Cover letter	Cover letter
	Research statement
	Teaching statement (for some)
	Your publications (if requested)
	Your references

If you do not already know the difference between a résumé and a CV, now is the time to learn. Use a CV when applying for faculty jobs in academic institutions and a résumé every place else. CVs are multipage and comprehensive, documenting your academic degrees, positions held, publications, awards and recognition, and teaching experience. Résumés are short, typically no more than two pages and often one page if you are an early career scientist. If you are applying for a position outside of the United States, these guidelines may differ. In the European Union, CVs may include personal details, language capabilities, and often a photo of yourself. In Asia (China, Japan, India),

CVs may contain much more detail including nationality or marital status.[29] Pay attention to country-specific expectations.

Use one of the very capable AI platforms to format and polish your résumé for you. As noted in Chapter 1, because of the ubiquity of AI, the time has long passed when hiring managers looking at a beautifully written cover letter and résumé conclude that the applicant has excellent language, writing, and organizational skills. You get no extra points for these things. On the other hand, precisely because of the ease of using AI, a poorly written cover letter with misspellings, and a long rambling résumé, are certain to confine your application to the reject pile.

There is no shame in using AI to format and polish your résumé, CV, and cover letters. Everyone else is doing it. If a prospective employer is really interested in your writing skills, they will have to find that out some other way. Indeed, the cynic may even wonder if writing skills should even be a criterion anymore, since even the poorest AI agent can take your disorganized thoughts and use them to write a treatise that reads like it was written by F. Scott Fitzgerald. Having said that, and as we will emphasize with the cover letter, the AI-generated version is a *starting place only*.

The table below shows the main elements of résumés and CVs. Since these may be scanned by a program that will search for key words matching the job description, they must be customized. Use words or phrases that reflect the specific position you are applying for. Do not cut and paste whole sentences or phrases as this will make your document look artificial. Insert your own language, and avoid jargon and any abbreviations that only a hyperspecialist in your subfield would understand.

Curriculum vitae (CV) and résumé compared

	CV	Résumé
Content guidelines	Must be customized for every position to use or reflect the key words or requirements of the job you are applying for. If you do not do this, your résumé or CV may get eliminated by their electronic scanning system.	
Length	As long as needed	1–2 pages maximum
	Major sections	
Begin each section with a descriptive title, in bold, as shown to the right ("**Education**," "**Positions held**," etc.) to make navigation simple.	• Name, mailing address, email, phone, current position title, and organization. • **Education**, reverse chronological order. • **Positions held**, reverse chronological order, years held, position title, organization, adviser or supervisor if a training position (i.e., grad student or postdoc). For each position, one or two accomplishment-centric (not activity-centric) bullets. Name the role or position, your responsibilities, one or two positive outcomes or accomplishments using descriptive action-oriented bullets for each (more detail below).	• Name, mailing address, email, phone, current position title, and organization. • **Profile** succinctly shows who you are and your skill set, strengths, and accomplishments. Your profile must reflect that you are a good fit for the job you are applying for. Two or three sentences. • **Career goal.** Include if you are applying for your first job or changing fields.

(Continued)

	CV	Résumé
	Customize these to highlight skills that are specifically relevant to the job you are applying for. • **Teaching experience** (more detail below). • **Skills.** This depends on your current position and what you are applying for. If it is a faculty position, leave this out. Your skills will be apparent from your publications and training. • **Awards, honors.** • **Presentations** made at major events, meetings, conferences. A seminar in the lab down the hall is not a presentation, but an invited seminar at another institution or even department is. • **Service.** Includes committee memberships, both local and national. If you held a leadership position or were responsible for initiatives, specify what they were and what you accomplished. • **Professional society memberships.** If you held a leadership position or were responsible for initiatives, specify what they were and what you accomplished. • **Publications**, in reverse chronological order. Include any that are "in press," as well as submitted (specify where). Do not include "in preparation" as no one will have any idea what this means.	• **Previous experience** or **Positions held.** For each position, one or two accomplishment-centric (not activity-centric) bullets. Name the role or position, your responsibilities, one or two positive outcomes or accomplishments using descriptive action-oriented bullets for each. Reverse chronological order. Customize these to highlight skills that are specifically relevant to the job you are applying for. • **Skills.** Depends on the position. If it is for a scientist job, this is a must. As above, highlight skills that are relevant to the position you are applying for. • **Education.** Reverse chronological order. • **Honors, awards.** • **Languages in which you are fluent.** May be important in a company that has an international presence.

When describing previous positions held, or roles you played in organizations, focus on accomplishments and outcomes, using action verbs rather than activities you engaged in. Examples are shown below. The weak words in the second column are in bold, and the strong action words in the third column are in bold and italics.

Strong words and weak words

Role	Weak: Vague and passive	Strong: Specific and active
Postdoctoral fellow in Cousteau lab.	**Studied** effects of ocean acidification on mollusk genetics.	***Discovered*** that acidified conditions triggered a 40% upregulation of shell-formation genes while suppressing immune response pathways.

(*Continued*)

Role	Weak: Vague and passive	Strong: Specific and active
Teaching assistant	**Ran** a discussion and lab section for Introduction to Oceanography.	***Led*** weekly 90-minute discussion sessions for 25 undergraduate students and supervised hands-on laboratory experiments analyzing water salinity and temperature data. Students ranked my sections in the top 10% of departmental sections.
Helped organize the university biotechnology club.	**I was one of two** founders and fundraisers for our university's biotechnology club.	***Cofounded*** the university's biotechnology club, ***recruited*** 45 charter members, and ***organized*** monthly industry speaker events that ***connected*** 4 students with internship opportunities at local biotech companies within the first academic year. ***Raised*** $10,000 from industry sponsors.

MAKING YOUR RÉSUMÉ AND CV SPEAK TO THEIR NEEDS

Your résumé should highlight your background and experience as they relate to the position description. While you should not exaggerate, you also should not downplay what you have done, something that scientists from cultures that value modesty may find difficult at first.[30] In the United States, it is common and expected that you will speak positively and enthusiastically about your skills and accomplishments.

Academia: For each position held, include key scientific accomplishments (just a few words) as well as activities that academic search committees will be interested in, such as mentoring and teaching experience, courses taught or developed, departmental or institutional responsibilities, and any awards or recognition for teaching, research, or service. You will elaborate on this in a separate teaching statement. Last will be your publications. There are often no page limits for academic CVs, but if one is specified adhere to it religiously. If you run out of space, you can always say something like "complete publication list available on request."

Here is an example of a weak entry in a CV to be used for an academic job and how it can be improved.

2023–2025. Postdoctoral Research Associate in lab of Dr. Famous Scientist, Stanford University. Conducted experiments in neural cell signaling. Published two papers, two more in preparation. Skills learned: cell culture, electrophysiology, cryo-electron microscopy.

What is wrong with this? First, no one cares if you "conducted experiments." As described above, this is a "weak" phrase. They care about what you achieved. Second,

"Skills learned" belong in a section entitled "Skills" if they ask for that; otherwise, your skills should be apparent from the description of your accomplishments. Published two papers? It is better to describe one or two key findings or results as accomplishments. Leave papers to the "Publications" section if you include one.

Here is a better version in which we have italicized the "strong" words used to describe this position.

> **2023–2025, Postdoctoral Research Associate, Stanford University, laboratory of Dr. Famous Scientist.** *Identified* a *novel* signaling pathway between hippocampal cells implicated in neuronal degeneration in Alzheimer's disease. *Created* screening methodologies, *first of their kind* knockout mice using CRISPR methods, *developed and applied novel* behavioral screens. This work *laid the groundwork* for my future studies in other functional brain regions. *Mentored* an undergraduate senior thesis project student as well as two summer students. *Helped create and taught* for three years a highly rated section of "Introduction to Neurobiology" summer course for local high school students.

For an academic job they are going to be interested in both your individual accomplishments as well as (and maybe more importantly) how your accomplishments translate into a viable research program that you will bring with you (this will be elaborated in your research statement). Since they are also interested in your mentoring and teaching experience, we added two sentences to this position to cover those skills. These could also be included in a section of your résumé entitled "Teaching and mentoring."

Other factors that can help your academic résumé are examples that illustrate your community spirit. Joining an academic institution means more than just building a lab and doing your science. It means being a member of a professional community that will need your active participation in activities that you may not see as directly related to your own advancement. Committee work, such as being on a graduate student admission committee, curriculum review or development committees, and grad student advisory committees, will be expected. Providing evidence that you are community minded and a team player at the organizational level by listing any committees or groups you have been a member of, especially in a leadership role, will be a big plus. Also be sure to mention how your research program will provide research and thesis opportunities for students and opportunities for collaboration with other faculty at the institution.

Industry: Unlike an academic CV, the résumé needs to be short and to the point. Do not include a publication list unless it is requested or you have extra space (in which case two or three publications may be all that is needed) or anything about teaching. As we suggested earlier, go back and pretend that you are the hiring manager and you are skimming your résumé to determine whether you should actually read it. Focus on what the *hiring manager* would want to know about you, not on what *you* want them to know about you. If it is not relevant to the job, do not include it or spend more than a few words on it. That does not mean that you should skip over listing a previous job or role; it means emphasizing your most relevant jobs or roles. Include brief descriptions about what you actually did or accomplished more than on techniques you know.

Here is the same poorly written entry that we used earlier, except now it is going to be in a résumé intended for an industry job application.

2023–2025. Postdoctoral Research Associate in lab of Dr. Famous Scientist, Stanford University. Conducted experiments in neural cell signaling. Published two papers, two more in preparation. Skills learned: cell culture, electrophysiology, cryo-electron microscopy.

Here is a better version for an industry job:

2023–2025, Postdoctoral Research Associate, Stanford University, laboratory of Dr. Famous Scientist. *Worked with a team* that used advanced genetic techniques and *developed* state-of-the-art behavioral screens to identify novel signaling pathways between hippocampal cells in mouse brains that are implicated in neuronal degeneration in Alzheimer's disease. *Hired and managed a technician* for this project. *Established* and served as *lead scientist in a currently ongoing collaboration* with two outside research teams to extend this work into other brain regions.

The words or phrases that convey actions and accomplishments as well as features valued by industry (teamwork, hiring, managing) are in italics. Scientists are often tempted to stress their independence and ingenuity in presenting themselves. That is fine in an academic setting, but in industry they want you to be a team player since all industry research is a team enterprise. Thus, the inclusion of the candidate's team and managerial skills.

Customizing your résumé has a dual benefit. First, it highlights those skills that are most relevant for the position you are applying for. Second, it shows the prospective employer that you know what is important to them and that you are not just mailing out generic résumés. You will learn more ways to customize your résumé or CV in the following sections.

MEET OUR JOB CANDIDATES

To illustrate how to create custom résumés, CVs, and cover letters, we will utilize two fictional candidates. The first is Samantha Jiminez, who we met in the previous chapter. Samantha is an analytical chemist who is going to apply for the Hyperbole Pharmaceuticals industry job we reviewed earlier. The second is Morgan Chen, a marine geneticist who is going to apply for the IMG marine genomics position.

Anthropic's AI agent Claude was used to create a résumé and CV. Claude was instructed to use the CV and résumé formats shown previously to create Samantha's résumé (Appendix 1) and Morgan's CV (Appendix 2). Claude was further instructed to create these documents so that each candidate would be a good but not perfect match for the requirements of the respective jobs they are applying for and to emphasize the areas where the résumé and CV matched the position requirements.

Claude's résumé and CV were carefully reviewed for accuracy, and awkward and artificial-sounding wording was removed or corrected. As is typical for current AI models, both looked good on the surface. But they had hallmarks of AI-generated text, including *weak words*, far *too many adjectives and adverbs*, and *phrases and word usages that were awkward* and in some cases *nonsensical statements* (more on this later). See the following examples:

- Claude's original effort contained the phrase "develop and validate robust analytical methods." No scientist on earth would have put the word "robust" in there, so we took it out.
- Removed "significantly" from the phrase "and significantly shortened development timelines."
- Eliminated the phrase "while maintaining a spotless safety record" because it sounded like puffery.
- Removed "Operated and maintained sophisticated analytical instruments while ensuring data integrity and traceability" since it sounds like it was copied from the position description.

Industry and academic-specific emphasis for each position held by either candidate was added as in the examples above. Claude could have been asked to do this, but writing the instructions would have taken longer than doing it myself. The resulting collaboratively created résumé and CV are well written, correctly formatted, of the correct length, and show that each of their backgrounds, training, and accomplishments mesh with what is required in the position descriptions. These are good examples of résumés and CVs that have been customized for the respective positions and can now be used as part of Samantha's and Morgan's application packages.

If you were applying for either of these positions and were appropriately qualified, there is no reason why you should not take your generic résumé or CV, along with the formatting and content guidelines for each, feed them to an AI agent along with the position description, and ask the agent to customize your résumé or CV for either position. Then, as done here, further customize each entry in your résumé or CV to highlight the key characteristics or features of each position held that are relevant for the job you are applying for. Carefully review *every word and phrase* of what you get back from your AI agent before you do anything further with it to remove any hint of the dreaded "AI voice." With your customized résumé or CV in hand, you are ready for the next step, preparing your cover letter.

CHAPTER 3 SUMMARY

Start much earlier than you think

- The job search process takes 8–14 months for average candidates in average markets, so plan backward from when you need to be gone and start well before your target exit date.

The golden rule: "It's not about you, it's about them"

- Focus on solving the employer's problems rather than what they can do for you. Customize every application to show how your skills meet their specific needs; do not just list your accomplishments.

Be brutally honest about fit

- Only apply for jobs you are genuinely qualified for and actually want. Get a trusted adviser to review job descriptions with you—if they squirm when you ask if you are qualified, do not apply.

Decode the hidden meanings in job postings

- Requirements are often wish lists where "essential" skills are more negotiable than they appear, while "preferred" qualifications are nice-to-haves that will not disqualify you.

Master the résumé vs. CV distinction and customize everything

- Use CVs for academic positions (comprehensive, multi-page) and résumés for industry (1–2 pages max). Both must be tailored for each application using keywords from the job posting and action-oriented language that emphasizes accomplishments over activities.

APPENDIX 1

Industry résumé

Samantha Jiminez
Analytical Scientist, Advanced Drug Delivery
127 Story St.
Cambridge, MA 02137
Phone: (762) 287-2837
Email: SamanthaJ572@gmail.com

PROFILE

Experienced analytical scientist with five years of industry expertise in analytical methods development for characterization of peptides, proteins, small molecules, and biologics. Proven track record in developing robust analytical methods using HPLC, UPLC, LC-MS, and SEC-MALS technologies for biopharmaceutical applications. Strong background in cross-functional collaboration, regulatory compliance (cGLP), and innovative analytical technology implementation. Demonstrated ability to support drug development programs from early discovery through clinical development with exceptional attention to detail and scientific rigor.

PROFESSIONAL EXPERIENCE

Senior Analytical Scientist | Pfizer | Cambridge, MA | 2020–Present

- Develop and validate analytical methods for quantification and characterization of peptides, proteins, and bioconjugates using HPLC, UPLC, LC-MS, and SEC-MALS
- Successfully implemented new mass spectrometry approaches that reduced characterization time for protein therapeutics by 35% while increasing sensitivity for detecting low-level impurities
- Led implementation of high-throughput LC-MS/MS platform that streamlined bioconjugate analysis and shortened development timelines
- Collaborate with formulation, process development, and quality teams to ensure drug candidates meet critical quality attributes
- Authored twelve technical reports and regulatory submissions following cGLP guidelines
- Mentored junior scientists on proper laboratory practices and instrument operation protocols

Analytical Scientist II | Moderna Therapeutics | Cambridge, MA | 2019–2020

- Developed analytical methods for biologics characterization using SEC-MALS, DLS, and HPLC techniques
- Established stability-indicating assays and method validation protocols for protein therapeutics
- Supported IND submissions with comprehensive analytical method validation packages

(Continued)

- Collaborated with process development teams to optimize analytical testing strategies for manufacturing scale-up
- Trained laboratory personnel on sophisticated analytical instruments and analytical procedures

Research Associate | Takeda Pharmaceuticals | Cambridge, MA | 2017–2019

- Supported small molecule and biologics drug development through method development and validation activities
- Executed analytical testing for drug substance and drug product characterization across entire analytical workflow
- Performed impurity profiling and identification using LC-MS techniques with focus on low-level impurity detection
- Contributed to regulatory filings and responded to health authority questions regarding analytical methods

EDUCATION

Master of Science in Analytical Chemistry | Northeastern University | Boston, MA | 2017

- Thesis: "Development of LC-MS/MS Methods for Quantification of Therapeutic Proteins in Biological Matrices"
- GPA: 3.8/4.0

Bachelor of Science in Biochemistry | University of Massachusetts Amherst | Amherst, MA | 2015

- Minor: Chemistry, Magna Cum Laude, Phi Beta Kappa
- GPA: 3.7/4.0

TECHNICAL SKILLS

Analytical Instrumentation: HPLC/UPLC, LC-MS/MS, LC-MS, SEC-MALS, DLS, CE-SDS, UV-Vis Spectrophotometry, FTIR, NMR

Software and Data Analysis: Empower, MassLynx, ChemStation, JMP, Prism, LIMS systems, Electronic Lab Notebooks

Laboratory Automation: Hamilton Microlab STAR, Tecan Freedom EVO liquid handling systems, high-throughput LC-MS/MS platforms

Regulatory Knowledge: GLP, ICH guidelines, FDA analytical method validation, regulatory requirements for drug development

Specialized Techniques: Method development and validation, biologics characterization, bioconjugate analysis, impurity profiling, stability testing, formulation characterization

SELECTED PUBLICATIONS

1. Jiminez, S., Johnson, K.L., Martinez, S.R. "Analytical Method Development for Characterization of Protein-Bioconjugate Formulations Using Multi-Angle Light Scattering." *Journal of Pharmaceutical Sciences*, 2022, 111(8), 2245–2254.

(Continued)

2. Thompson, R.A., Jiminez, S., Chen, M.H. "Stability-Indicating HPLC Method for Simultaneous Quantification of Therapeutic Proteins and Degradation Products in Biologic Drug Development." *Pharmaceutical Research*, 2021, 38(12), 2018–2031.
3. Jiminez, S., Rodriguez, P.M., Kumar, A.S. "Implementation of High-Throughput LC-MS/MS Platform for Bioconjugate Analysis in Biopharmaceutical Applications." *LC-GC North America*, 2020, 38(11), 567–575.
4. Davis, L.K., Jiminez, S., Williams, J.T. "LC-MS/MS Method Validation for Quantification of Protein Therapeutics Following ICH Guidelines and cGMP Requirements." *Bioanalysis*, 2019, 11(18), 1687–1702.

PROFESSIONAL AFFILIATIONS AND CERTIFICATIONS

- American Chemical Society (ACS)—Member since 2017
- American Association of Pharmaceutical Scientists (AAPS)—Member since 2018
- Certificate in Biopharmaceutical Analysis—University of California San Diego Extension (2019)

AWARDS AND RECOGNITION

- Pfizer Pharmaceuticals Excellence in Innovation Award for Mass Spectrometry Implementation (2023)
- Moderna Therapeutics Outstanding Contribution to Analytical Development (2020)
- Graduate Research Excellence Award, Northeastern University (2017)

APPENDIX 2

Academic CV

Morgan Chen, Ph.D.
Department of Marine Biology
University of California, Santa Barbara
129 Helix Way
Santa Barbara, CA 98765
Email: mchen134@ucsb.edu | Phone: (675) 763-2378
ORCID: 0000-0002-1234-5678

CURRENT POSITION

NSF Postdoctoral Research Fellow (2022–Present)
Laboratory of Dr. Jacques Cousteau
University of California, Santa Barbara, Department of Marine Biology
Project: Genomic Adaptation to Ocean Acidification in Marine Keystone Species

EDUCATION

Ph.D. in Marine Biology (2022)
Stanford University, Hopkins Marine Station, Pacific Grove, CA
Dissertation: "Evolutionary Genomics of pH Tolerance in Marine Invertebrates"
Adviser: Dr. Betty Crocker
GPA: 3.89/4.0

M.S. in Marine Sciences (2018)
University of Washington, Seattle, WA
Thesis: "Population Genetics of Intertidal Communities Under Environmental Stress"
Adviser: Dr. Michael Rodriguez
GPA: 3.92/4.0

B.S. in Biology, Magna Cum Laude (2016)
University of California, San Diego, La Jolla, CA
Concentration: Marine Biology and Ecology
Minor: Environmental Chemistry
GPA: 3.85/4.0, Phi Beta Kappa

RESEARCH EXPERIENCE: **Marine Evolutionary Genomics | Ocean Acidification Research | Adaptive Responses**

NSF Postdoctoral Research Fellow (2022–Present)
University of California, Santa Barbara, Jacques Cousteau Lab

- Led multidisciplinary research characterizing adaptive genomic responses to ocean acidification in three keystone marine species
- Discovered novel carbonic anhydrase gene variants conferring enhanced pH buffering capacity in Pacific oyster populations

(Continued)

- Collaborated with oceanographers, biochemists, and climate modelers on integrative research approaches
- Mentored four undergraduate summer students and two graduate students

Graduate Research Assistant (2018–2022)
Stanford University, Hopkins Marine Station, Betty Crocker Lab

- Investigated evolutionary mechanisms of pH tolerance in marine invertebrates using comparative genomics
- Developed novel bioinformatics pipelines for population genomic analysis of non-model organisms
- Conducted field research at multiple Pacific Coast sites from Baja California to Alaska

Research Technician (2016–2018)
Friday Harbor Laboratories, University of Washington, Oscar Meyer Lab

- Supported research on climate change impacts on marine ecosystems
- Maintained live marine organism cultures and conducted physiological experiments
- Assisted with molecular laboratory techniques including DNA extraction, PCR, and sequencing

---

TEACHING EXPERIENCE

Teaching Assistant (2022–Present)
University of California, Santa Barbara

- Coastal Ecology Course (Dr. Oscar Meyer): Led weekly laboratory sections for twenty-four students, designed field exercises, graded assignments (three years)
- Marine Genomics Course (Dr. Sarah Lee): Taught bioinformatics workshops, mentored student research projects, developed new curriculum modules (three years)

Guest Lecturer (2021–2022)
Stanford University

- "Climate Change and Marine Adaptation" (Undergraduate Marine Biology)
- "Genomic Tools in Conservation Biology" (Graduate Seminar Series)

Undergraduate Mentor (2019–2022)
Stanford University

- Supervised two undergraduate honor thesis projects in marine genomics
- Coauthored three peer-reviewed publications with undergraduate mentees

---

SELECTED PUBLICATIONS **(complete list attached)**

Chen, M., Thompson, K.L., Rodriguez, P.A., & Cousteau, J. (2024). Carbonic anhydrase gene variants drive local adaptation to ocean acidification in Pacific oyster populations. *Nature Climate Change*, 14(8), 892–901. DOI: 10.1038/s41558-024-02045-x

(*Continued*)

Martinez, S.R., **Chen, M.**, Kim, D.H., & Lee, S. (2024). Comparative transcriptomics reveals species-specific responses to pH stress in calcifying marine organisms. *Global Change Biology*, 30(3), 1245–1262. DOI: 10.1111/gcb.17089

Chen, M., Crocker, E.N., Johnson, A.B., & Smith, C.D. (2023). Genome-wide association mapping identifies pH tolerance loci in the purple sea urchin *Strongylocentrotus purpuratus*. *Molecular Ecology*, 32(15), 4123–4139. DOI: 10.1111/mec.16985

GRANTS AND FELLOWSHIPS

National Science Foundation Postdoctoral Research Fellowship in Biology (2022–2025)
"Genomic mechanisms of rapid adaptation to ocean acidification"
Award #: DBI-2205789 | $216,000

Stanford Graduate Fellowship (2018–2020)
Full tuition and stipend support | $45,000/year

PADI Foundation Grant (2021)
"Community-based monitoring of coral reef resilience"
$15,000

UC San Diego Undergraduate Research Scholarship (2015–2016)
$5,000

AWARDS AND HONORS

- Outstanding Postdoctoral Researcher Award, UCSB Marine Science Institute (2024)
- Best Student Presentation, International Symposium on Ocean Acidification (2021)
- Graduate Student Research Award, Western Society of Naturalists (2020)
- Phi Beta Kappa Honor Society (2016)
- Dean's List, UC San Diego (2014–2016)

PROFESSIONAL SERVICE

Peer Review Activities
Nature Climate Change, *Global Change Biology*, *Molecular Ecology*, *Marine Biology*, *PLOS ONE* (2022–Present)

Conference Organization
Program Committee Member, Pacific Coast Ocean Acidification Symposium (2023–Present)

Professional Memberships

- American Society of Naturalists (2018–Present)
- Western Society of Naturalists (2019–Present)
- Society for Molecular Biology and Evolution (2020–Present)
- Ocean Acidification Research Community (2021–Present)

(Continued)

SELECTED PRESENTATIONS

Invited Talks

Chen, M. "Genomic adaptation to ocean acidification: Lessons from Pacific marine invertebrates." Scripps Institution of Oceanography Seminar Series, La Jolla, CA. March 2024.

Chen, M. "Rapid evolution in changing oceans: Molecular mechanisms of pH tolerance." University of Oregon Marine Biology Colloquium, Eugene, OR. November 2023.

Conference Presentations

Chen, M., Rodriguez, P.A., & Cousteau, J. "Carbonic anhydrase variants as a mechanism for local adaptation to ocean acidification." Ocean Sciences Meeting, New Orleans, LA. February 2024. (Oral presentation)

Chen, M., Kim, D.H., & Lee, S. "Transcriptomic signatures of pH stress across marine calcifiers." International Symposium on Ocean Acidification, virtual conference. September 2023. (Oral presentation)

Chen, M., Thompson, K.L., & Crocker, E.N. "Population genomics reveals rapid adaptation to environmental pH gradients." Western Society of Naturalists Annual Meeting, Monterey, CA. November 2022. (Poster presentation)

TECHNICAL SKILLS

Laboratory Techniques: DNA/RNA extraction, PCR, qPCR, library preparation for sequencing, protein expression and purification, enzyme assays, marine organism husbandry

Computational Skills: Bioinformatics pipeline development, population genomics analysis, transcriptome assembly, phylogenetic analysis, statistical modeling in R, Python programming, high-performance computing

Field Experience: Subtidal and intertidal sampling, SCUBA diving (PADI Advanced Open Water), water chemistry analysis, organismal physiology experiments

REFERENCES

Dr. Jacques Cousteau
Professor and Director, Marine Science Institute
University of California, Santa Barbara
Email: cousteau@ucsb.edu | Phone: (805) 555-0100

Dr. Betty Crocker
Professor of Marine Biology
Stanford University, Hopkins Marine Station
Email: bcrocker@stanford.edu | Phone: (831) 555-0150

Dr. Sarah Lee
Associate Professor of Marine Genomics
University of California, Santa Barbara
Email: slee@ucsb.edu | Phone: (805) 555-0175

Dr. Oscar Meyer
Professor of Oceanography
University of Washington, Seattle
Email: omeyer987@uw.edu | Phone: (206) 555-0200

4 The Craft of Cover Letters

Industry vs. Academia (and Why AI Won't Save You)

PLAN TO USE AI? PROCEED WITH CAUTION

Once Samantha and Morgan have read and deciphered the position descriptions, determined that they want to apply for their respective jobs, and customized their résumé or CV, their next step is to write a cover letter. As always, their goal is to *show* their prospective employers how well they match the requirements in the position posting.

The cover letter is the first place where they get to tell the hiring committee or manager who they are. Different jobs will require that they highlight different facets of their background, and each application letter will reflect that. The cover letter is the one element that both academic and industry applications have in common, but as you will see, the content and format of these are different. The principal goal of an industry cover letter is to show that the candidate has the background and skills to do the job. Academic cover letters are more complex. They need to highlight the candidate's scientific accomplishments, show that they have a promising research program, and say something about their teaching and mentoring skills.

As we did with the résumé and CV, we are going to use Claude to generate a cover letter for each candidate. But unlike with the résumé and CV, which required only straightforward editing, the cover letters are a very different story. Using an AI-generated cover letter just as it is is asking for trouble.

The analytical chemist position description and Samantha's résumé were fed into Claude with the prompt "Write a cover letter for Samantha to apply for this position" with no further instructions. Of course, you would start with *your* résumé, feed it to Claude with the position description, and ask Claude to write your letter highlighting how your résumé matches the position description. Sounds like a snap, right? Not so fast.

Despite Samantha being a candidate who matches the qualifications in the job description pretty well, the limitations of the AI cover letter generated by my simple prompt were significant. This is the cover letter for Samantha that Claude generated.

I am writing to express my interest in the Analytical Chemist position at Hyperbole Pharmaceuticals. With my extensive background in analytical chemistry specializing in biologics and small molecule characterization and method development, I am eager to contribute my expertise to your innovative drug development programs.

Throughout my career, I have developed and validated robust analytical methods for the quantification and characterization of peptides, proteins, and bioconjugates using advanced techniques including HPLC, UPLC, LCMS, and SEC-MALS. My experience encompasses the entire analytical workflow from method development through validation and troubleshooting. At Pfizer, I successfully implemented new mass spectrometry approaches that reduced characterization time for protein therapeutics by 35% while increasing sensitivity for detecting low-level impurities.

I have consistently demonstrated proficiency in evaluating emerging analytical technologies and integrating them into existing workflows. For example, I led the implementation of a high-throughput LC-MS/MS platform that streamlined bioconjugate analysis and significantly enhanced our development timeline. My technical expertise is complemented by strong collaborative skills, having worked effectively with formulation, process development, and quality teams to ensure drug candidates meet critical quality attributes.

My approach to laboratory work emphasizes both scientific rigor and safety. I have maintained a spotless safety record while operating and maintaining sophisticated analytical instruments, and I've trained junior scientists on proper laboratory practices and instrument operation protocols. I am well-versed in electronic laboratory notebooks and meticulously document all experimental work to ensure data integrity and traceability.

With my bachelor's degree in Biochemistry and master's degree in Analytical Chemistry, combined with five years of industry experience in biopharmaceutical analysis, I believe I am well-positioned to make immediate contributions to your analytical team. I am particularly excited about the opportunity to apply my knowledge of cGMP and regulatory requirements to support your development programs.

I welcome the opportunity to discuss how my skills and experience align with your needs. Thank you for considering my application, and I look forward to speaking with you soon.

My earlier advice to use AI to help write your letter did not mean you should use it as is. Even our pretty good candidate Samantha needs to think twice about sending this letter. The letter looks pretty slick, just what you would expect from AI. The hiring manager may think "Is this a real person? If they are trying to fool me in their cover letter, will they also try to fool me when they have to update me on their project that's not going so well?"

What are the giveaways in this letter? For starters, certain phrases just sound too fancy.

- "Throughout my career": If you have been at it for fifteen years, fine, but if it has only been five years it sounds like puffery.
- "the entire analytical workflow": Really? The *entire* workflow? You most likely only worked on some small part of it.
- "reduced characterization time for protein therapeutics by 35% while increasing sensitivity": If you did this, you would probably be getting a promotion, not looking for a new job.

- "consistently demonstrated proficiency": Nobody talks like this.
- "significantly enhanced our development timeline": Business speak, not for a scientist.
- "strong collaborative skills": Sounds like you copied this from the position posting.
- "critical quality attributes": Nobody talks like this.

Before Samantha gives up and writes her own letter, let's not abandon AI quite yet. After all, if this is not what she wants, she could just give Claude some additional guidance and ask for a better letter. So, Claude was given this prompt: "The following is a job application letter generated by AI (here I inserted the above letter). It sounds too slick to me, not like it was written by a scientist. Make the letter less slick and more like a scientist would really write." Here is what came back:

I'm writing to apply for the Analytical Chemist position at Hyperbole Pharmaceuticals. Given my background in analytical chemistry with a focus on biologics and small molecule characterization and method development, I believe I could contribute effectively to your drug development programs.

During my work experience, I've developed and validated analytical methods for quantifying and characterizing peptides, proteins, and bioconjugates. The techniques I regularly employ include HPLC, UPLC, LCMS, and SEC-MALS. My work has covered method development, validation protocols, and troubleshooting instrument issues. While at Pfizer, I implemented several mass spectrometry approaches that reduced protein therapeutic characterization time by approximately 35% and improved detection sensitivity for low-abundance impurities.

I've spent considerable time evaluating new analytical technologies and figuring out how to incorporate them into our existing lab workflows. One project involved setting up an LC-MS/MS platform that made our bioconjugate analysis more efficient and helped accelerate development timelines. Beyond the technical aspects, I've worked closely with colleagues in formulation, process development, and quality control to ensure our drug candidates meet the required quality specifications.

In the lab, I prioritize experimental accuracy and safety protocols. I've maintained good safety practices while using and maintaining analytical instruments and have trained newer lab members on proper instrument operation. I'm comfortable using electronic lab notebooks and am meticulous about experimental documentation to maintain data integrity.

I hold a bachelor's degree in Biochemistry and a master's in Analytical Chemistry, plus five years working in biopharmaceutical analysis. I'm familiar with cGMP requirements and regulatory guidelines relevant to analytical work in drug development.

I'd be interested in discussing how my experience might fit your team's needs. Thank you for reviewing my application.

This is better. First, it is shorter, which is good. Second, the letter has lost a lot of the modifiers (*strong*, *entire*, *consistently*, *effectively*) so it sounds a bit more realistic. But it still needs considerable editing. Some phrases are either awkward ("*contribute effectively*"), too colloquial ("*figuring out*"), or just sound silly ("*prioritize experimental accuracy*")—what does that even mean? She does some inaccurate experiments, but they are not a high priority? This makes no sense. So, she is still going to have to do a lot of careful reading and editing.

Another reason to use AI with caution is that an AI-generated cover letter may inadvertently lead naive employers to conclude that the candidate has skills that they do not have. For example, an employer not yet attuned to the world of AI may use cover letters to assess language and writing skills. An AI-generated letter obviously subverts this goal.

If the hiring manager concludes that Samantha has stellar writing, organizational, and communication skills when she does not, and if they do not pick up on this anywhere else (i.e., the interview and reference checks), then they may end up hiring someone who really does not have the skills they need. This could be bad for the hiring manager and bad for Samantha. They are going to feel duped, and Samantha is going to be doing a job at which she may fail. But it is the employer's fault if excellent writing skills were not listed in the position description, and there is not much Samantha could have done to prevent this misperception. Below you will see how to deal with job requirements that are spelled out but that you do not have.

WRITING A BETTER INDUSTRY COVER LETTER

These are the principles to keep in mind as you create your letter:

- Your goal is to convince them that you have the skills they need for their projects.
- Highlight that you are a team player and have good interpersonal presence and skills.
- Keep it short. This is much harder than rambling, but it is essential for you to have any chance of getting noticed. Never exceed a page and a half, and your goal should be one page.

The letters generated by Claude for Samantha show the danger of using AI output without careful instruction, scrutiny, and editing. To take full advantage of AI in writing, your cover letter you must do the following:

1. Provide the AI agent with all the relevant data. In the case of the industry cover letter, that includes your résumé and the position description.
2. Carefully instruct the AI agent on the desired format and ideal length for the output—we did not do that in Samantha's case but will do it in the following.
3. Have the AI agent generate a first draft, review it carefully, and correct awkward phrasing, silly statements, flat-out mistakes, lies, and tone that does not sound like a scientist.
4. Refine instructions to the AI agent if specific items need emphasis or alteration.
5. Generate a second draft and do Step 4 again if needed.

By the time you are done, you may wonder if you shouldn't have just written the letter yourself. But in either case, you need to know how a good cover letter is structured. Here is an outline for the content and layout of an industry cover letter.

Industry cover letter content

Letter section	Purpose	Comments
1. Your name and contact information	Your current institution	This shows where you are now.
2. Salutation		
3. Introduction	State the position you are applying for and where you saw the posting. State who you are and where you work now. Give a one-sentence summary of your skills and training. State your interest in this job and your admiration for the company.	This paragraph is a short preview of the rest of the letter.
4. Your background and scientific skills	One or two short paragraphs on your training, skills, and experience.	Describe your training and background. Detail your technical skills or qualifications, aligning them one by one with those stated in the position description. Use some of the key words from the position description. List one or two quantifiable achievements if possible.
5. Your interest in this company and job	Why you are interested in this specific job and this specific company. Show that you know something about them. Show your enthusiasm for the job and restate why you are a good fit.	This shows them that this is not a generic letter that you send to every company.
6. Closing paragraph	Restate your enthusiasm and interest in the job, indicate your hope that you can meet with them at their convenience.	Express enthusiasm for a meeting.
7. Sign-off and signature		Insert jpg signature.

One-page industry cover letter: What it looks like on the page

1. Your name and contact information
2. Salutation
3. Introduction
4. Your background and scientific skills
5. Your interest in this company and job
6. Closing: Restate your enthusiasm for the job, tell them you look forward to hearing from them
7. Sign-off and signature

The hiring manager, HR, or people on the search committee are pressed for time. They will scan your résumé to see if you meet their basic criteria (if this has not already been done for them by an AI bot or someone in HR, in which case they may never see your résumé at all). Then they will scan your cover letter. Consciously or unconsciously, they will be matching you to who they are looking for. If you have done an excellent job customizing your letter to highlight how you meet their criteria, and this is backed up by your résumé, you may make it to the next step.

Appendix 1 contains Samantha's final cover letter for the analytical chemist job. We generated this by giving Claude the position description and Samantha's résumé as input and also told Claude to generate a one-and-a-half-page cover letter using the above format. We then edited Claude's output as we discussed to remove silly, awkward phrases and incorrect information. The letter is annotated, showing exactly which parts of the letter make reference to which parts of the position description. This letter should get the appropriate attention whether it is scanned by a bot or read by a human.

These are the major parts of the letter:

Introduction. After the salutation, state the name of the position you are applying for and where you saw or heard about it and one or two sentences describing who you are (be brief; your résumé will include the details). As in the rest of the letter, whatever you write here needs to address or reference something that is in the position description. Here are some examples:

- ***From the position description.*** Utilize subject matter knowledge in analytical chemistry to develop and verify analytical methods to quantify and characterize peptides, proteins, small molecules, and drug substances in a variety of formulations.
- ***How Samantha responded.*** I am currently a Senior Analytical Chemist at Pfizer with an M.S. in Chemistry from Northeastern University and five years of experience in pharmaceutical analytical chemistry. My expertise encompasses advanced method development for proteins, peptides, and small molecules using LC-MS, UPLC, and SEC-MALS systems.

Another, shorter option is:

> I am a highly experienced analytical chemist with a strong track record of scientific accomplishments in the field of peptide chemistry and characterization.

Someone else applying for a different job might say:

> I am a molecular geneticist nearing completion of a four-year postdoctoral position in the laboratory of Dr. Very Famous.

Background and scientific skills. While your résumé lists your training and skills, here you have an opportunity to highlight how specific aspects align with the job description. Samantha does a great job of this, as might be expected. In her letter in Appendix 1 we see how almost every sentence in this section of her letter refers to some element of the job description. Samantha also uses some of the key words from the description and includes one quantifiable achievement.

- ***From the position description.*** General knowledge of cGMP, GLP (cGMP and GLP refer to current good manufacturing practice, and good laboratory practice, respectively), and regulatory requirements for pharmaceutical industries are a plus.
- ***How Samantha responded.*** "… including two Phase II support programs done under GLP compliance. The latter included compiling the analytical data packages for two regulatory submissions …"
- ***She could also have used a shorter version.*** "I am responsible for generating data collected under GLP compliance for regulatory submissions."

While Samantha does have GLP experience, she has *no* cGMP experience. Taking the description at face value, anyone can develop a "general knowledge" of almost anything with just a couple of hours on the Internet. So, after a couple of hours she could honestly have said yes to "general knowledge of cGMP." What if it read instead "two to three years of working in a cGMP environment required." That is a higher standard, and she either meets it or she does not. On the other hand, say she has most or all of the other requirements. Now, not having cGMP experience does not seem like the kiss of death, and we have already told you that it is fine to apply for positions for which you do not have every one of the stated requirements.

As noted in Chapter 3, the way she should handle this is not to mention it in her letter. The letter should talk about what she *can* do, not what she *cannot* do. It is for them to decide how important that skill is, especially since this particular element may simply be a "plus" for them rather than being a gating element, no matter what they said in the position description. If it is important to them, it will come out in the interview. You will see how to handle that in the next chapter.

It is also possible that the current candidate pool is full of people with marginal cGMP qualifications. This means that the hiring manager is going to have to make some compromises, which could come in the form of relaxing their requirements. If this happens, and if she is the one whose application checks the most boxes, in addition to convincing them that she is a quick learner, the cGMP item could assume a lower importance.

Further comparison of the qualifications section with Samantha's experience will also show that she does not have any or much experience in "advanced drug delivery technologies" but, as with the cGMP, she will not mention this.

Your interest in this company and job. Show them you know who they are and why you want this job. This section needs to convey that you are familiar with the institution, company, organization, or lab and why you want to work there. Samantha says:

> I am particularly drawn to Hyperbole's commitment to innovative therapeutic solutions and advanced drug delivery technologies.

She could also say:

> I am a great admirer of Hyperbole Pharmaceuticals' programs, specifically your ability to identify novel approaches to the treatment of rare diseases such as TDF disorder.

Closing paragraph. Here you express additional enthusiasm for the job, say you think you are a great fit and that you admire the company, and you would really like to meet them in person.

WRITING AN ACADEMIC COVER LETTER

If you skipped the previous section on writing an industry cover letter, go back and read it regarding using AI to help (or hinder) you, then come back.

An academic job application is more complex than one for industry. In addition to your cover letter and CV, you will need to prepare one or more supporting documents, the most important of which is a description of your proposed research program. You also may need to prepare a teaching statement. We will cover each of these below.

Your academic letter will be read with one of the same objectives as an industry letter: They are going to be looking for any way to eliminate your CV from the pile as quickly as possible. So, you need to cover every or most items in the position announcement clearly and concisely. As with an industry position, you need to show that you are who they are looking for in about one and a half pages. Since they will also be looking at your CV, do not include publications, degrees, etc., in your letter. *The cover letter allows you to add "color commentary" to your application by saying things that do not appear anywhere else in your application materials.*

Academic cover letter content

Letter section	Purpose	Comments
1. Name and contact information		
2. Salutation		
3. Introduction	State the position you are applying for and where you saw the posting. Review your background and scientific achievements. State your area of research focus and what is unique or notable about how you approach your work. Review your scientific strengths (special or unique skills are key), past achievements (discoveries, publications, grants), and training. Express enthusiasm.	The goal is to grab the reader with something interesting or unusual that makes them want to read on. Summarize a few key accomplishments that demonstrate your scientific expertise relevant for this position, your productivity, and training. Mentioning your training here establishes your credentials early in the letter, especially if you trained in a recognized place or with a leader in your field, as can showing that you were productive—numbers of publications and/or awards or grants.
4. Research program	The big questions that your current and future research program seeks to address. Especially for an R1 institution, you need to show that you plan to attack and advance the most important questions and issues in your field. You need to convey the message that your program will yield high impact publications and independent funding and that your presence will address the current needs of the department (expressed in the position posting) and provide growth or expansion in areas that may not be currently represented there.	For R1 and R2 institutions, this is the most important part of the letter. It gives them a quick read on whether your work will fit with the department and more importantly whether it will be fundable. Even for teaching-focused positions, having a strong and clear research agenda is critical to show that you are experienced and expert in your field.

(*Continued*)

Letter section	Purpose	Comments
5. Teaching	If relevant for this position, a few sentences about your experience as well as interest in being a teacher and mentor and how your research provides opportunities for students to do research in your lab. Demonstrate your credentials and interest as an educator.	This will be most important in institutions in which teaching is a major element of the job. For most R1 institutions, teaching is not a major factor, and if mentioned in the job posting it will be a secondary consideration. It is important to pay close attention to wording in the job ad—if teaching is highlighted in the ad, make sure to highlight it in your cover letter.
6. Fit in department	A sentence or two about how your interests complement those of others in the department. The goal is to show that your presence could lead to productive collaborations or the establishment of critical mass expertise in areas that they value.	You need to show them how your proposed program aligns with and expands their focus or mission.
7. Closing and more enthusiasm	You can't show too much enthusiasm.	
8. Name and signature		Insert jpg signature.

Academic cover letter: What it looks like on the page

First page
1. Name and contact information
2. Salutation
3. Introduction
4. Research program

Second page
Research program continued
5. Teaching, research opportunities for students if relevant
6. Fit in department
7. Closing and more enthusiasm
8. Name and jpg signature

The people reading your letter are looking for "surrogate markers" for your expected success as a researcher and faculty member. These include where you trained, who you trained under, your publication record, your research field, if you have funding, and if you do not, how likely it will be that you will get funding. If you tick all or four out of five of these boxes, you may get a screening interview.

Above are layout and content guidelines for a cover letter for an academic faculty position. Using this layout, the reader will know where to look to get the information they need quickly. While you may think you want to stand out from the crowd and get noticed, the format of the cover letter is not the place to do that. Search committee members looking through dozens and dozens of CVs want to be able to jump to key places in your letter to get a quick read on whether you are a viable candidate and whether your CV is worth spending more than five seconds on. They will look at where you trained, if they know or know of your adviser, what field you are in, and what you accomplished as a postdoc. Having the proper format helps them do that. Trying to stand out by formatting your letter as a graphic novel will not be a smart move.

Appendix 2 contains an application cover letter for the Institute for Marine Genomics position shown in the previous chapter for our fictional candidate Morgan Chen. As we did with Samantha, we created Morgan's CV to be a pretty good but not perfect match for the position description. Also, as we did for Samantha, we used Morgan's CV and the position description as input and told Claude to write the application letter according to the guidelines and format. As we did with Samantha's AI-generated industry letter, we carefully proofread and edited the first AI draft to eliminate mistakes, awkward phrasing, and errors and then customized the entries to reflect the needs expressed in the job description. Morgan's letter is about a page and a half. In Appendix 2 the text of the letter is in the left column with the sections of the letter labeled as described earlier. To the right of each section is a commentary describing the key features of the section and the approach used in writing it.

The letter makes the case that Morgan is an excellent match for the position. This of course will not be the case for every candidate, and we are just using Morgan to show you what an excellent match looks like. As discussed above, real candidates will emphasize the areas of overlap in their experience and plans with the position description and skip over areas that do not overlap.

CHAPTER 4 SUMMARY

AI-generated cover letters must be carefully edited

- They often sound too slick and artificial. Hiring managers can spot AI writing and may question your authenticity or worry you are hiding poor communication skills.

Industry cover letters must be concise and team-focused

- Keep it to one page, emphasize being a team player with collaborative skills, and directly match your qualifications to their specific requirements using keywords from the job posting.

Academic cover letters are longer and research-focused

- Use one and a half pages to highlight your scientific achievements, describe your future research program and big questions you will tackle, and show how you complement existing faculty and can attract funding.

Customization is nonnegotiable for both industry and academia

- Every letter must be tailored to the specific position description, using some of their language and addressing their stated needs rather than sending generic applications.

Structure matters more than creativity

- Both industry and academic letters should follow established formats that make it easy for busy hiring managers to quickly find key information—this is not the place to "stand out" with unusual formatting.

APPENDIX 1

Industry cover letter with annotations

1. Name and contact information Samantha Jiminez 127 Story St. Cambridge, MA 02137 Phone: (762) 287-2837 Email: SamanthaJ572@gmail.com June 25, 2025	**Specific elements of Essential Functions and Qualifications from the position description are reproduced and numbered below to show where they are referenced in the letter.**
2. Salutation Hiring Manager (or their name if specified) Hyperbole Pharmaceuticals 121 First St. Cambridge, MA 02346 Dear Hiring Manager:	
Section headings (in blue) and numbers in parentheses are not part of the letter. The numbers in parentheses refer to the numbered functions and qualifications taken directly from the position description, excerpted and shown in the right-hand column.	
3. Introduction I am writing in response to your advertisement for the position of Analytical Chemist, posted on Monster.com. I am currently a Senior Analytical Chemist at Pfizer with an M.S. in Chemistry from Northeastern University (1) and five years of experience in pharmaceutical analytical chemistry (2). My expertise encompasses advanced method development for proteins, peptides, and small molecules (3) using LC-MS, UPLC, and SEC-MALS systems (4). I am excited about the opportunity to contribute to Hyperbole's therapeutic development programs.	(1) Master's degree within subject matter expertise can be considered (2) 4+ years' relevant experience + pharmaceutical industry knowledge (3) Quantify and characterize peptides, proteins, and small molecules (4) HPLC, UPLC, LCMS, SEC-MALs, MS instrumental analysis techniques

(Continued)

4. Background and scientific skills My master's thesis, "Development of LC-MS/MS Methods for Quantification of Therapeutic Proteins in Biological Matrices," gave me a strong foundation in analytical methods used in biologics formulation (5). At Pfizer, I developed and validated over fifteen analytical methods for drug substances in a variety of formulations (6) and played a key role on the team that introduced robotic technologies that increased our analytical throughput by 40% and reduced development timelines by an average of three weeks per project (7).	(5) Method development for biologics formulations (6) Develop and verify analytical methods to quantify and characterize. . . drug substances in a variety of formulations (7) Develop analytical technologies and methods to increase current analytical capabilities + evaluate and implement new analytical technologies
I am fully certified in operating, maintaining, and troubleshooting Waters UPLC systems, Thermo mass spectrometers, and Agilent GC-MS platforms (8). As Senior Analytical Scientist, I developed a novel MS approach for glycosylated peptide conjugates (9) and managed three interdisciplinary development projects (10) including two Phase II support programs done under GLP compliance (11). The latter included compiling the analytical data packages for two regulatory submissions (12). My team has worked extensively with the formulations group to help validate drug substance properties (13), and I am recognized as an effective communicator who can translate complex scientific concepts for diverse audiences (14).	(8) Operate, maintain, and troubleshoot analytical lab instruments including HPLCs, UPLCs, and MS (9) Advanced separation and detection methodologies (10) Project management and working collaboratively with cross-functional teams (11) General knowledge of cGMP, GLP, and regulatory requirements for pharmaceutical industries (12) Generate product supporting data packages (13) Collaborate with in-house teams to define and validate dosage and chemical attributes of the drug candidates (14) Excellent scientific communication skills and ability to engage with stakeholders effectively + strong attention to detail and ability to document experimental work electronically
5. Your interest in this company and job I am particularly drawn to Hyperbole's commitment to innovative therapeutic solutions and advanced drug delivery technologies (15) and strongly believe that my track record of implementing novel technical solutions to improve workflow will enable me to help advance your programs (16). Your novel project management model, incorporating interdisciplinary project coleads, is one that resonates strongly with my own project management experience (17).	(15) Advanced drug delivery technologies (16) Develop analytical technologies and methods to increase current analytical capabilities (17) Project management and working collaboratively with cross-functional teams

(Continued)

6. Closing I am genuinely excited about joining Hyperbole's analytical chemistry team and contributing to the success of your therapeutic programs. I would welcome the opportunity to discuss how my methods development expertise (18), regulatory compliance experience (19), and focus on adopting innovative analytical approaches can help support your company's mission.	(18) Strong knowledge and expertise in analytical chemistry, including method development (19) General knowledge of cGMP, GLP, and regulatory requirements
7. Sign-off and signature Sincerely, *Samantha Jiminez* Samantha Jiminez	

APPENDIX 2

Academic cover letter with annotations

Section headings (in blue) are not part of the letter.	
1. Your name and contact information **Morgan Chen, Ph.D.** Department of Marine Biology University of California, Santa Barbara 129 Helix Way Santa Barbara, CA 98765 Email: mchen134@ucsb.edu Phone: (675) 763-2378 June 23, 2025	Current institutional letterhead and contact information.
2. Salutation Search Committee Institute for Marine Genomics Cornell University Ithaca, NY Dear Search Committee Members:	
3. Introduction I am writing to apply for the position of Assistant Professor at the Institute for Marine Genomics at Cornell University (IMG), which was advertised in the June 1 issue of *Science*. I am currently completing my third year as an NSF-funded postdoctoral fellow in the laboratory of Jacques Cousteau at the University of California, Santa Barbara. During this time, I have led a multidisciplinary effort that has characterized adaptive genomic responses in three keystone marine species to ocean acidification. My work led to the discovery of a suite of carbonic anhydrase gene variants that confer enhanced pH buffering capacity in certain populations of the Pacific oyster (*Crassostrea gigas*). I am especially interested in this position for the opportunities it provides to interact and collaborate with the outstanding faculty and students at IMG.	Highlight right away your academic credentials and well-known mentor. Use the words "multidisciplinary" and "collaborative" right away since that is something they clearly signaled they want. Mention a "discovery" (strong word) as opposed to "worked on."

(*Continued*)

4. Research program The long-range objective of my program is to transform global marine conservation by providing the first predictive framework for identifying which ocean ecosystems will survive climate change, enabling proactive protection of genetically resilient populations before they collapse. The genomic tools developed will have the potential to revolutionize the marine aquaculture industry by enabling breeding of climate-adapted species, ensuring food security for coastal populations as traditional fisheries decline under ocean acidification. By establishing the genetic foundations of ecosystem resilience, this work would create a new paradigm for climate adaptation science that could be applied globally across all marine and terrestrial ecosystems, fundamentally changing how humanity prepares for and responds to environmental change. Specifically, my research program will extend carbonic anhydrase mechanism discovery across multiple keystone calcifying species including mussels, sea urchins, corals, and coralline algae, using comparative phylogenomics and ecosystem-level eDNA surveys to map acid-tolerance alleles across marine communities from high-variability environments. CRISPR-based functional genomics will be used to validate the causality between genetic variants and physiological resilience, while investigating how pH tolerance genes interact with responses to climate stressors. The program will integrate population genomic data with climate projections to create predictive maps of ecosystem vulnerability and develop algorithms that assess resilience based on keystone species' genetic composition, while establishing breeding programs and seed banks for climate-resilient marine organisms. Finally, the research would produce field-deployable genetic assays for rapid resilience assessment and establish a global collaboration network to standardize methodologies and create an international database of marine climate-adaptation genomics.	You need to grab them here or they will not go any further. Starting with a long-range objective that seeks to "transform" anything is as good as it gets. As above, avoid "studied," "undertook," and "worked on," which sound like you are telling them how you spent your time, not what you accomplished. Attention-grabbing words and phrases ("first," "survive," "proactive," "revolutionize," and "new paradigm") may sound overly grand, but you are talking about your vision, not your list of specific aims. About halfway through, signaled by the term "specifically," the statement moves onto a list of plans that you could also have prefaced with "in my first three years. . . and during the next two years" to demonstrate how these will phase in. It ends with an aspirational goal to create an "international" database. Everyone loves databases.

(Continued)

5. Teaching I have been a teaching assistant at UCSB for three years, teaching sections and labs in Dr. Oscar Meyer's Coastal Ecology course and in Dr. Sarah Lee's popular Marine Genomics course. My teaching experiences have been among the most satisfactory aspects of my training at UCSB, and I am ready to take responsibility for teaching these or other courses that you may need at IMG. I anticipate that my research program will offer ample opportunities for independent projects for both graduates and undergraduates.	Since the posting mentions teaching, you need to address that here. This looks like an R1 institution, so teaching skills or experience may not be in their top three criteria for hiring you, but better safe than sorry. It cannot hurt to say you enjoy teaching.
6. Fit in department Finally, I am especially attracted to this position as it offers exciting opportunities to work with existing faculty at IMG. I believe that Dr. Ron McDonald's widely recognized work on coral bleaching, and Dr. Howard Johnson's innovative studies of ocean biogeochemistry, could offer exciting opportunities for collaboration and for future multidisciplinary grant applications. I look forward to meeting them both.	The job post mentions or alludes to collaboration twice, which is their way of saying it is important. Morgan had already signaled his collaborative nature earlier in the letter, and here is his opportunity to get more specific. Morgan has been a bit bold and mentioned two faculty members, praising their work. In cases where collaboration is not so prominently mentioned, it is best not to get this specific.
7. Closing and more enthusiasm I am attaching my CV, a two-page statement of my research interests and vision for my program, and a one-page teaching statement, as well as the names and contact information of three references as requested. I am excited about this opportunity and hope to meet the members of the committee in person. I look forward to hearing from you.	
8. Name and signature Sincerely, *Morgan Chen* Morgan Chen	Congratulations, you followed their instructions.

5 The Academic Trifecta

Research Vision, Teaching Philosophy, and Strategic References

THE ACADEMIC RESEARCH STATEMENT

If you are applying for an academic position in a research institution, your application package will need to contain a research statement, considered by many to be the most important part of your application.[1] Positions in industry do not ask for such statements because you will be working on their projects, not yours.

Institutions may provide their own detailed instructions and guidelines for research statements, but a general rule is that your statement should be one to three pages long. The research statement will typically be the first place the search committee will see your research plans and focus. If the position advertisement states that applicants in a specific field of research are being sought, then your research statement needs to reflect that.

Your research statement needs to spell out your vision for your research, your key contributions to the field, and a description of your proposed research program.[31] Use short declarative sentences and be concise. Long complex sentences will induce glazed eyes and skimming. Never exceed the space limit or they may not read your statement at all. Have your research statement read in advance by as many people as you can. Ask them for their frank feedback on its clarity and organization.

While you should begin your research statement with a few sentences that will be understandable to nonexperts in your field, the bulk of the statement will be targeted to the one or two actual experts in your field who will read all of it. The nonexperts will be influenced more by your publication record and the people you list as references and what they say about you than by how your work fits into and extends a field they are only peripherally familiar with.

While you should target experts with your statement, in your job talk and one-on-one interviews you must be able to talk about your research in a way that the nonexperts will understand. Since many scientists struggle with this without expensive coaching and input from friendly colleagues, having others listen to your talk is an absolute necessity. Even if you think you are good at explaining your work to nonspecialists, this is simply a hypothesis, and you need to test it.

As with your résumé and cover letter, the research statement needs to be customized for each institution, emphasizing aspects of your research that appeal to their needs. That means choosing what to talk about and what to save for later (i.e., your job talk).

Here is a layout for a typical research statement.

Academic research statement
Approximate length: 2 pages (10-point font, single spaced)

First page	Second page
Summary: 25% of a page—15 lines	Plans for near and intermediate term (3–5 years): 50% of a page—25 lines
Background work: 40% of a page—20 lines	Long term directions and significance: 30% of a page—20 lines
Current projects: 25% of a page—15 lines	

The elements you need to convey in a research statement include the following:

- The important problem you are working on
- Your previous significant contributions
- Your research plans
- How your focus fits with what they are looking for
- Your research will attract funding
- You are a thoughtful mentor and excellent teacher (briefly, as this will be expanded in your teaching statement)

Research statement content

Label all sections according to the institution's instructions so they can jump to what they are interested in.

1. **Summary:** ***one paragraph.*** A well-crafted summary is a hook that should compel the reader who is only planning to read the summary and nothing else to continue reading. Your message needs to be succinct and well-crafted.

    A. The big picture—what problem you are addressing and why it is interesting and important

    B. One or two specific aims you hope to accomplish in your first three years

    C. Your aims and your unique approach to the problem

D. One piece of exciting preliminary data
E. One sentence on future directions
F. One sentence on how this work fits into the big unanswered questions of your field

2. **Background:** about a ***half-page.*** This and the following two sections show the arc of your research program—the important scientific problem you are addressing, what is known about this problem, what is unknown, and how you plan to fill that gap. Briefly describe a couple of your notable accomplishments and how they advanced this field. It is OK to include one interesting graphic or figure. Even though you are targeting specialists, try to make the figure easily understood by the nonspecialists. Assume that any figure you already have does not meet this criterion and invite one or more nonspecialists to look at it to help you make it do so.
3. **Current projects:** ***half-page.*** List only a couple. Describe the big questions they seek to answer and why they are exciting.
4. **Plans for near and intermediate term:** about a ***half to one page.*** Plans for the first year and the first three to five years. About a third to half of your statement should be focused on the future. Include two or three specific aims for the first three years with an outline of your approach or strategy. Mention how your program fits with, and expands on, what else is going on in the department.
5. **Long-term directions and significance:** about a ***third of a page.*** They are looking for programs that will have longevity. This section needs to show your vision for your research program and how it extends the boundaries of your scientific field.

I asked Claude to use the above format to create a two-page research statement consistent with the brief research description in Morgan Chen's application letter that we wrote earlier. I carefully edited the statement as described earlier for the cover letters. The result is in Appendix 1. For illustrative purposes I labeled each section of the statement according to the above scheme and annotated each paragraph to show how it addresses the specific requirements of that section.

Your approach will be similar except that you will start by writing your own research statement using the above format, doing your best to emphasize areas that match what the position description calls for. Then you will feed your research description, the layout guidelines, and the position description into the AI agent and ask it to improve upon what you have written to make the areas of overlap more apparent and the statement better organized. If you start by asking the AI agent to write your statement for you, rather than edit it, you will spend the rest of the day editing out mistakes and exaggerations.

Go over the resulting research statement word by word to make sure that this is in fact your program and that the language being used is yours and not that of the AI agent. Your research statement is the foundation for much of what you will be discussing with and presenting to the search committee. It will also be the basis of your chalk talk, some of your research seminar, and probing questions from the few faculty members who are experts in this exact area of research. It must be yours and sound like you.

THE ACADEMIC TEACHING STATEMENT

An academic position often comes with the expectation that you will teach. How much and how often will depend on the type of institution you are applying to. Some of this will be negotiable during the hiring process. We will get into that in Chapter 8.

Being articulate about your approach to teaching, referred to as your *teaching philosophy*, and your teaching experience in the form of a teaching statement is a part of the application package that may feel alien to some scientists. If you are fortunate enough to have been in a Ph.D. program in which you were either required or able to be a teaching assistant as part of your training, you may also have had some exposure to pedagogy via short training courses. Not having such training is not necessarily a deal-breaker but having it is definitely a plus. Various sources indicate that at least a quarter of STEM Ph.D. recipients have had some teaching experience, but the actual number may be as much as twice that.[32,33]

Even if the only teaching you have ever done was as a teaching assistant in a course, that may be enough. One survey found that among 317 applicants for faculty positions in 2018–2019 (67% were in R1 institutions), those whose only teaching experience was as a teaching assistant (TA) were no less likely to get academic job offers than those with more extensive teaching experience.[1]

If you have not been a TA, think broadly about anything you have ever done that could remotely qualify as teaching. Have you ever mentored someone, even in a nonacademic setting? Did you train a technician or undergraduate to perform a lab technique? Have you been a camp counselor or held a summer job where you trained others to perform their tasks? Were you in a club or social organization where you led some activity? All of these can be conveyed as teaching experience. An added benefit is if you can speak extemporaneously about enjoying teaching, how you approached it, and how you adapted your teaching style to the needs of those you taught.

If you have been a TA at an institution that had a training program for its teaching assistants, you may have a head start in being able to discuss various approaches to teaching, how to engage students, how to help students with special needs, and possibly how to draft a curriculum plan. This is good training since if you get the position, you will likely have to discuss your or others' teaching styles when evaluating other faculty or during self-evaluations for tenure decisions. Even so, your teaching statement will likely be the first time you will be writing this topic.

A sample of learning and teaching concepts that could be included as part of a teaching philosophy is shown in the following list. Do some reading about these or other approaches and pick a couple that resonate with you. You can get additional ideas by reading sample teaching statements that are widely available online.[34,35]

- Active learning—how active participation and in-class activities promote learning.
- The value of realistic lab and classroom activities. Having students do real-world experiments that either are or mimic how science is actually done promotes engagement and identification with the subject matter. For example, "Students will use publicly

available genetic data on BLAST to reconstruct evolutionary relationships between vertebrate taxa."

- Aligning teaching with assignments, lab work, class discussions, and evaluation so that they are all focused on conveying a small group of core competencies, rather than a disconnected group of facts.
- Highlighting the power of collaboration. Creating student project teams can teach the power of collective problem solving and prepare students for how much science is done in collaboration with others.
- Harnessing the power of feedback. Feedback needs to be an integral part of a teaching philosophy. Giving positive feedback reinforces adaptive behaviors, and providing constructive feedback in a nonpunitive manner shows students how they can do better.

Since everyone knows you are not a philosopher of education, they will correctly assume that the high-minded ideas you convey about teaching are not yours. No one expects you to invent your own teaching philosophy any more than they expect you to have invented your own branch of mathematics. As you read about various educational approaches and look at sample teaching statements, identify a couple of approaches that make the most sense to you and that resonate with your own experience. Prepare a couple of anecdotes that illustrate how you use those approaches to engage with students or trainees.

A typical teaching statement has the following elements:

1. ***Your teaching philosophy***—how you think about and approach teaching. Do not try to come off as a professor of education. Speak from personal experience as much as possible. Avoid dogma and categorical statements (saying "Students learn very little by sitting in lectures" could make the inveterate lecturers in your committee nervous). If you reference a specific teaching approach, have an example of what you are talking about from your own experience. Without this, someone may ask "What do you mean by student-centered teaching?" and you will be at a loss because it was just something you read about.
2. ***Your teaching experience.*** Be specific: "Responsible for weekly problem review sessions with 10–15 students that involved recapping the key learnings from the lectures and helping individuals with specific learning needs." A short but salient anecdote that illustrates a teaching experience is a big plus. "As a lab TA for Biochemistry 101, I was responsible for making sure the weekly lab exercises were set up properly. During labs I showed students how to troubleshoot the assigned protocols and helped them relate what they were doing in the lab to what they learned in class. The 'ah hah' moments I encountered were many—I especially recall one situation where a student made a mistake in the buffer composition in an SDS gel sample that caused the sample not to enter the gel. This morphed into a wide-ranging group discussion of the nature of detergents that got everyone engaged in a very immediate manner." If you have received student evaluations on your teaching, what were they and what did you learn from them? Do not be overly concerned if you had less than stellar ratings

as long as you can talk about how you used this feedback to improve your teaching. Frame this as your having a "growth mentality" when it comes to teaching.

3. ***How your teaching approach enables students to learn and how you assess learning.*** The best teaching statements focus on student outcomes—what you want them to learn and how you go about facilitating and assessing that.
4. ***What you can teach.*** Be explicit about what you can teach and how that matches the needs and curriculum of this department. Research their current course catalogue and let them know which courses you would be prepared to teach.
5. ***Courses you could develop.*** If you have heard that they have topical gaps in their curriculum, consider suggesting how you might be able to fill one of them. Suggest one or two new courses you could develop (but not in your first year) that they are not currently offering. If the job posting mentions specific courses that need to be taught by a new hire, make sure to specifically address your qualifications and interest in teaching those courses.

Here is an infographic showing how the above elements fit into the teaching statement, taken from the MIT Communication Lab website:[36]

Academic teaching statement anatomy (one page)

1. Your teaching philosophy. 10 lines, 20%
2. Teaching experience and approaches used. 10 lines, 20%
3. How you assess learning. 10 lines, 20%
4. Existing courses you could teach. 10 lines, 20%
5. Courses you could develop. 10 lines, 20%

Other issues that you need to think about, especially if you are applying to an institution that is a primarily undergraduate institution (PUI), include the following:

- What is your strategy with students who are struggling with the subject matter?
- How would you describe your classroom style?
- Describe a challenging situation you encountered while teaching. How did you manage that?
- If you do not have a lot of teaching experience, be up front about your desire and willingness to learn.
- Describe a personal learning experience that was transformative or eye opening for you. Telling such stories during an interview can be a powerful way to show your desire to be a passionate and compelling teacher.

If most or all of your teaching experience has been at an R1 institution, consider this advice from Phoebe Cohen (a professor of geoscience at Williams College):

A good teaching statement also includes something specific about the institution—this is especially important if your teaching experience is at an R1, but you are applying for a position at a liberal arts college, for example. We want to know you have done research on things like class sizes, class formats, instructional lab facilities, etc. In addition, I look for evidence of a growth mindset—many applicants will only have TA experience or limited solo classroom teaching experience—I want to see that they know they have a lot to learn and that they have a plan or a mindset to grow as teachers.

We have included a teaching statement for our hypothetical marine geneticist Morgan Chen in Appendix 2.

QUEUE UP YOUR REFERENCES

Some places will ask for a list of references to be sent with your application letter, and others may ask for these only if you are selected for the next step, which will be a screening interview. In either case, make sure you have the list ready and that you have the permission of your references to send their names. Choose these people carefully.

Unless there is a good reason, you must include your current or previous adviser or boss. Not including them may be a red flag, suggesting in the worst case that you are an underperformer. There may be circumstances where you do not want to or cannot include them, and if that is the case, be prepared to explain why. For example, your current boss may not want you to leave because they see you as essential to *their* continued success and want to keep you for as long as possible without regard for *your* best interests. In such a situation, you may not want them to know you are looking to leave out of legitimate concern that this will negatively affect your working relationship with them.

Unfortunately, this is a hard story to tell even when it is true. Be cautious when explaining it that you do not come across as disparaging of your boss or that you sound disgruntled. These can be red flags for a prospective employer. If you are convinced that you cannot use your current boss as a reference, you will need to explain it in as neutral a manner as possible. For example, "Right now it would be challenging for me to let my adviser know I want to leave. They would like me to stay, but I think it's time for me to move on professionally. I completely understand their viewpoint, but I need to do what's best for me right now. They've been a great adviser, and I've learned a lot from them scientifically. I'm hoping that my other references can give you an accurate picture of my strengths." Hopefully, the hiring committee will get the message.

It is awkward to ask potential references whether they will say good things about you. Hopefully you know which ones are going to be most positive. On the other hand, there is nothing wrong with asking "I'm wondering if you'd feel comfortable writing a letter of recommendation for me for this job?" Let them know the job or jobs you are applying for so that they can customize their letter for that job.

Let your references know well in advance—at least a month—when you will need their letters. If you are applying for more than one or two positions at a time, send them a spreadsheet with the relevant information for each position, including the date their letter is due, and with links to each job advertisement or position description. Make sure that you have given your references the *exact version* of your résumé or CV that you plan to send in for the job you are asking them to recommend you for. You do not want recommendation letters saying things about your background or accomplishments that are at odds with what is in your résumé or CV. Understandably, most will not customize their letter because it is a lot of work, but you may have one or two gems on your list who will. Remember that these people are busy and that your request may be number 67 on their to-do list. As the deadline for letters of recommendation nears (maybe two weeks out), gently remind them of your request and the due date.

WELCOME TO THE BLACK HOLE

Once you have sent in your application package, it is time to sit back and await a stream of electronic and paper communications from the target organizations updating you on their progress at every step. Just kidding! Sadly, in nearly every case you will hear nothing unless you are chosen for an interview. One survey found that nearly 60% of academic job applicants received no feedback or rejection letter in response to submitted applications.[1] Even if you do hear that you were not selected for further action, you will most certainly not be told why. Search committees are overwhelmed with applications, and they simply do not have the time or staff to respond to all applicants, and even if they do respond, it is impossible to do so with any degree of customization. Do not despair. Understanding that their lack of response has nothing to do with you will help you maintain a positive attitude toward what is universally felt to be a time-consuming and stressful undertaking (more on this in Chapter 14).

CHAPTER 5 SUMMARY

The research statement is your most critical document

- This 1–3-page document should target the few experts in your field who will read it, clearly outlining the important problem you are addressing, your unique approach, preliminary data, and how your program will attract funding and fit with departmental needs.

Teaching statements require creative thinking about experience

- Many scientists lack formal teaching experience, so think broadly about mentoring, training technicians, being a TA, or any leadership roles. Research educational philosophies that resonate with you, and prepare specific anecdotes that illustrate your teaching approach.

Your references should include your current adviser (usually)

- Unless there is a compelling reason not to, you need your current boss/adviser as a reference. If you cannot include them, prepare a diplomatic explanation that does not disparage them while asking other references to provide a complete picture of your strengths.

Structure matters for research statements

- Follow the five-section format: compelling summary with hook, background showing your contributions, current projects addressing big questions, detailed 3–5-year plans, and long-term vision that demonstrates program longevity and field impact.

Expect radio silence after applying

- Nearly 60% of academic job applicants receive no feedback or rejection letters. The lack of response reflects institutions being overwhelmed with applications, not the quality of your materials, so try to maintain a positive attitude throughout the process.

APPENDIX 1

Academic research statement

(Underlines indicate key elements and are not included in the submitted statement.)

Research Statement: Foundations of Marine Climate Resilience

Summary

Climate change threatens marine ecosystems globally, yet we lack predictive tools to identify which populations will survive acidifying oceans—forcing conservation efforts to react to collapse rather than prevent it. My research program addresses this critical gap by developing the first genomic framework for predicting ecosystem resilience to ocean acidification, focusing on carbonic anhydrase mechanisms that enable survival in high CO_2 environments.

a. The big picture—what big problem you are addressing and why it is interesting or important

Over the next three years, I will map acid-tolerance genes across keystone calcifying species and integrate population genomics with climate projections to create predictive vulnerability maps for marine communities.

b. One or two specific aims you hope to accomplish in your first three years

My unique approach combines comparative phylogenomics of naturally acid-tolerant populations with CRISPR-based functional validation and ecosystem-level eDNA surveys to identify the genetic determinants of resilience before ecosystems collapse.

c. Your unique approach to the problem and aims

Preliminary work has identified novel carbonic anhydrase variants in mussels from naturally acidic environments that confer 40% greater survival under projected 2050 pH conditions compared to variants from stable-pH populations.

d. One piece of exciting preliminary data

This foundation will expand to establish breeding programs for climate-resilient marine species and develop field-deployable genetic assays for rapid ecosystem assessment.

e. One sentence on future directions

This work addresses the fundamental question of how genetic diversity drives ecosystem stability under environmental change, potentially revolutionizing our approach to climate adaptation across all marine and terrestrial systems.

f. One sentence on how this work fits into the big unanswered questions of your field

(Continued)

Background

The arc of your research program—important scientific problem, what is known, what is unknown, and how you plan to fill the void.

Ocean acidification represents one of the most pressing threats to marine biodiversity, with pH levels dropping 30% since the Industrial Revolution and projected to decrease another 150% by 2100. Calcifying organisms—including mollusks, echinoderms, and coral reef builders—face particular vulnerability as acidic conditions impair shell and skeleton formation, threatening entire ecosystem foundations. While previous research has documented widespread physiological impacts and species-level responses to acidification, a critical knowledge gap remains in understanding the genetic mechanisms that enable some populations to thrive in naturally acidic environments while others collapse.

What is known about the problem; establishing its importance

Recent advances in marine genomics have revealed that acid tolerance often involves carbonic anhydrase enzymes, which regulate internal pH by controlling CO_2-bicarbonate equilibrium. However, most studies focus on single species under laboratory conditions, missing the complex genetic interactions that determine ecosystem-level resilience. My previous work has demonstrated that naturally occurring genetic variants of carbonic anhydrase genes show remarkable diversity across marine populations, with some variants providing substantial protection against acidification stress. These findings suggest that the genetic tools for survival already exist within natural populations—we simply lack the framework to identify and protect them before ecosystems collapse.

Notable accomplishments and how they advanced the field, including your previous work

The field urgently needs a predictive approach that can identify resilient genotypes across multiple keystone species simultaneously, enabling proactive conservation rather than reactive management. By establishing the genetic foundations of acid tolerance across marine communities, this research addresses a fundamental gap in climate adaptation science while providing practical tools for ecosystem management.

What is unknown—the knowledge gap and how your work fills this gap

(Continued)

Current projects

List only a couple. Describe big questions they seek to answer and why they are exciting.

My current research focuses on two interconnected projects that establish the foundation for predictive marine resilience genomics. The first project involves comprehensive genomic analysis of blue mussel (*Mytilus edulis*) populations from naturally acidic environments, including CO_2 seeps and upwelling zones, compared to populations from stable-pH environments. Using whole-genome sequencing and transcriptomics, I have identified candidate acid-tolerance genes, with carbonic anhydrase variants showing the strongest association with survival under acidification stress. Functional validation using CRISPR-Cas9 gene editing in mussel embryos has confirmed that specific allelic variants confer up to 40% greater survival under projected 2050 pH conditions.

First project

Significance

The second project expands this approach to sea urchins (*Strongylocentrotus purpuratus*), examining how acid-tolerance genes interact with thermal stress responses as ocean warming co-occurs with acidification. Preliminary results indicate that genetic variants providing acid tolerance may compromise thermal resilience, revealing critical trade-offs that will determine which populations survive combined climate stressors. This work demonstrates the necessity of examining multiple stressor interactions when predicting ecosystem responses to climate change.

Second project

Significance

Plans for near and intermediate term

About half of the statement should focus on the future. Plans for first year and first 3–5 years; 2–3 specific aims for first 3 years with approach/strategy. Describe how the program fits with and expands what is in the department.

Year 1 objectives: I will establish comparative phylogenomic analysis across four keystone calcifying species (mussels, sea urchins, corals, and coralline algae) to identify conserved acid-tolerance mechanisms and species-specific adaptations. Using existing tissue collections from natural acid-tolerance hotspots, I will generate reference genomes and conduct population-level sequencing to map genetic variation in carbonic anhydrase and related pH-regulation pathways. Simultaneously, I will develop standardized CRISPR protocols

(*Continued*)

for functional validation across these taxonomically diverse species.

Years 2–3 objectives: Building on genomic foundations, I will conduct ecosystem-level eDNA surveys to map the distribution of acid-tolerance alleles across marine communities, linking genetic composition to environmental pH variability. This work will integrate population genomic data with high-resolution climate projections to create predictive maps of ecosystem vulnerability at regional scales. Collaboration with IMG's oceanography and climate modeling groups will be essential for incorporating physical oceanographic data into genetic predictions.

How program fits with department—collaboration

Years 3–5 objectives: The program will transition toward applied conservation outcomes by establishing breeding programs for genetically resilient marine organisms and developing field-deployable genetic assays for rapid ecosystem assessment. Working with the university's aquaculture facilities and marine stations, I will demonstrate proof-of-concept for genetic enhancement of commercially important species. This research directly complements the department's existing strengths in marine ecology and conservation biology while establishing new capabilities in conservation genomics that will attract additional funding and collaboration opportunities.

How program expands departmental capabilities, attracts resources, and will attract funding

More on departmental integration and collaboration

The integration of this program with departmental research is particularly strong given existing expertise in marine ecosystem dynamics and climate change impacts. Collaboration with faculty in ocean biogeochemistry will enhance the environmental context of genetic analyses, while partnerships with marine policy researchers will ensure that genetic tools translate into actionable conservation strategies.

Further elaboration on departmental integration

Long-term directions

Show that programs will have longevity. Vision for research program and how it extends boundaries of scientific field.

The ultimate vision for this research program extends far beyond marine acidification to establish genomic resilience assessment as a standard tool for ecosystem management under climate change. Over the next decade, I envision expanding the predictive framework to terrestrial ecosystems, examining how genetic diversity in keystone species determines community stability under drought, temperature extremes, and changing precipitation patterns.

Vision for extending field boundaries—from marine to all ecosystems

(Continued)

The methodological advances in functional genomics and ecosystem-level genetic monitoring will create broadly applicable tools for conservation biology.

Broad applicability showing field transformation

International collaboration will be central to long-term success, with plans to establish a global network for marine climate-adaptation genomics that standardizes methodologies and creates shared databases of resilience-associated genetic variants. This network will enable real-time monitoring of genetic erosion in threatened ecosystems and facilitate rapid response conservation actions.

Program longevity through international collaboration and infrastructure

The program will also pioneer the development of "genetic early warning systems" that can predict ecosystem collapse before traditional ecological indicators show decline.

Innovation showing field extension—predictive capabilities

Ultimately, this research program aims to fundamentally transform how humanity prepares for environmental change—shifting from reactive damage control to proactive genetic stewardship that maintains ecosystem function under novel climate conditions. By establishing the scientific foundation for predictive resilience assessment, this work will create new paradigms for conservation practice that could prevent the collapse of critical ecosystems globally, ensuring both biodiversity preservation and human food security in an uncertain climate future.

Ultimate vision for paradigm shift and global impact

APPENDIX 2

Academic teaching statement

Dr. Morgan Chen, Ph.D.
Marine Biology | University of California, Santa Barbara

TEACHING PHILOSOPHY

My approach to teaching is to help students understand fundamental scientific concepts by applying theory and observation to real-world problems. My classroom is a place where students apply critical thinking, observational, and analytical skills to pressing global issues like climate change and ocean acidification. By integrating current research findings, including my own work on genomic adaptation, I bring students into an ongoing scientific dialogue that will define our planet's future. I am skilled at creating a learning environment that accommodates diverse learning styles and ensures that all students can develop confidence in their scientific abilities.

TEACHING EXPERIENCE AND APPROACHES

Over three years as a teaching assistant at UC Santa Barbara, I have developed a classroom approach that emphasizes active learning and field-based experiences. In Dr. Oscar Meyer's Coastal Ecology course, I transformed traditional lab exercises into inquiry-based investigations where students developed hypotheses and designed experiments. My most successful innovation has been "ecosystem detective" activities where students analyze real environmental data from ongoing research projects, correlating water chemistry with species distribution patterns.

In Dr. Sarah Lee's Marine Genomics course, I introduced the use of bioinformatics tools and developed workshops where students analyzed genomic data from marine organisms using R and MatLab. These workshops culminated in student-led research projects presented in a mini-symposium format. My field-based approach includes tide pool expeditions and visits to aquaculture facilities, allowing students to observe organisms and processes discussed in class while developing research skills.

ASSESSMENT OF STUDENT LEARNING

I use assessment strategies that emphasize real-time observation and feedback. In laboratory settings, I use practical assessments where students practice and learn organism identification, equipment use, and data interpretation. For genomics coursework, I developed portfolio-based assessments where students compile bioinformatics analyses throughout the semester with written explanations.

I am a big fan of peer learning through collaborative projects in which groups explore complex problems that mimic real-world research programs. I provide group feedback and assessments at biweekly check-ins and provide individual students with detailed comments on their learning trajectory as well.

(*Continued*)

COURSES I COULD TEACH

Existing Courses:

- Marine Biology (Introductory): Comprehensive coverage from molecular to ecosystem levels, emphasizing current research and climate change impacts through extensive field work
- Conservation Biology: Integration of evolutionary adaptation research with traditional conservation topics like population genetics and habitat management

New Courses I Could Develop:

- Climate Change Biology: Adaptation and Evolution in Marine Systems: Upper-level course exploring organism responses to environmental change through genomic analysis and collaborative research projects
- Applied Marine Genomics: Interdisciplinary course teaching bioinformatics techniques with partnerships to local biotechnology companies

STUDENT TEACHING EVALUATIONS

My teaching evaluations consistently reflect strong engagement and learning outcomes. In Coastal Ecology, I maintained a 4.7/5.0 average over three years. In Marine Genomics workshops, I received 4.8/5.0 ratings. Representative student comments include:

> *"Morgan made ecology come alive through field trips and hands-on activities. Made me consider marine biology as a career path."*

> *"Learning to analyze real research data made me feel like a real scientist. These skills will definitely help in graduate school."*

6 Dancing on Eggshells

The Guide to High-Stakes Academic and Industry Interviews

MORE HOMEWORK: KNOW THEM AND THEIR NEEDS

If you are successful and your application letter and résumé get the attention of the hiring manager, recruiter, or committee, you will end up on a list of those to be screened by phone or video. The figures below show you the slightly different sequences of events from this point onward for both academic and industry jobs. While the sequence for an industry job is a bit simpler (see Chapter 11), almost everything in this chapter applies to both industry and academia.

Whether you are preparing for a screening call or a full-blown multi-day on-site interview, you need to find out who you will be meeting with and as much as you can about them. Even though most of a thirty-minute screening interview will consist of them asking you questions, you need to have two or three questions for them if time permits.

For example, in an academic screening interview, commenting on a recent publication by one of the search committee members is a good way to make a quick connection. In a screening call for an industry job with a recruiter or a member of the HR staff, you probably do not want to ask about the company's science. On the other hand, if it is with the hiring manager who is a scientist, having one or two questions about a recent scientific announcement made by the company could be a good move.

Create a one-page document for each person you will be meeting or speaking with in screening interviews and on-site visits. Include their photo, bullet points of salient information about them, and one or two questions for them specific to their research, position, or role. Online searching may be all you need to find someone's biography, LinkedIn profile, faculty web page, publication record, or social media accounts. Have these documents spread out in front of you during a video screening interview and bring them with you for an on-site interview. You may not get the chance to ask

Academic job application sequence

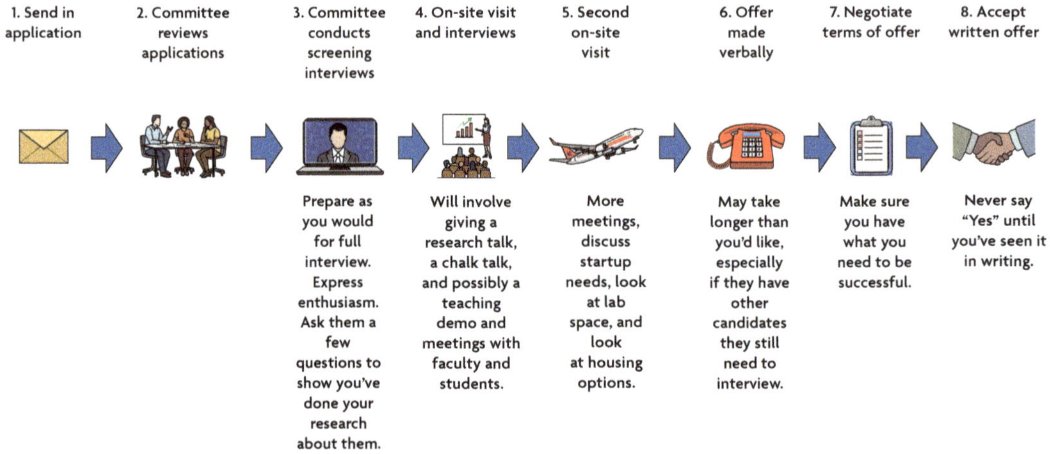

Industry job application sequence

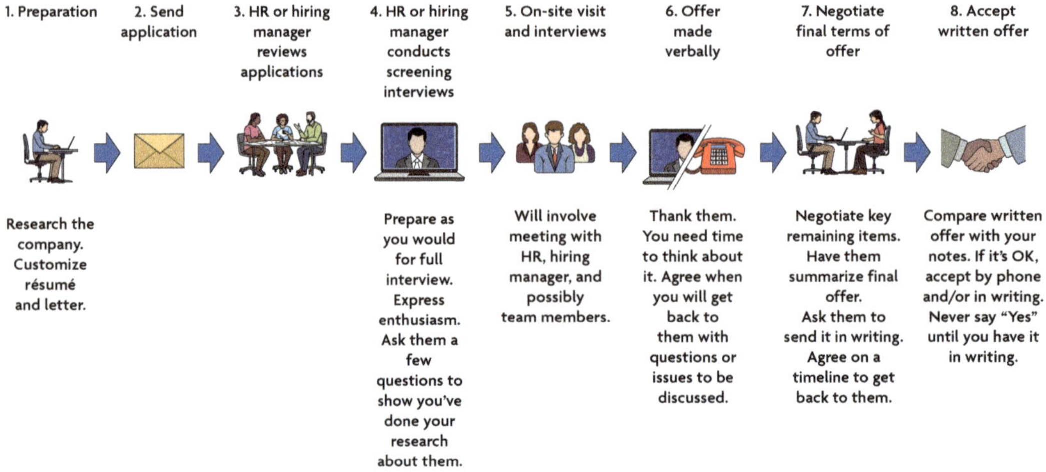

interviewer-specific questions in a screening interview, but you should be prepared to if they ask you what questions you have for them. Especially for an on-site interview, prepare thoughtful questions for each person you will be meeting. Even if you do not get to ask questions, making even the briefest comment about the job or a publication of the person interviewing you will indicate that you have done your homework and may be enough for you to stand out.

SCREENING INTERVIEWS: WING THESE AT YOUR PERIL

Depending on the organization, there could be as few as one and as many as a dozen or more people present. Their sole objective will be to determine whether you get an in-person interview. When they call you to schedule this interview, they may not refer to it as a screening interview. But if someone reaches out to set up a twenty- to thirty-minute phone or video call you should figure that is what it is.

There is a lot at stake for you in the relatively short time of the screening interview, so do not minimize its importance. This is not a rehearsal for a full interview. If you do not make it through the screening interview, you are not going to get a second chance. You need to prepare for it as you would for a full-blown face-to-face interview. Do not fool yourself into thinking that because it is only thirty minutes that you can wing it.

One facet of screening interviews that is almost always different from the full interview is the format. While the full-blown interview is nearly always in person, screening interviews are nearly always by video and sometimes by phone. Depending on whether your screening interview is by video or phone, there are specific strategies you can use to maximize your effectiveness.

Phone screening interviews

These are becoming rare, but on the off chance that you get a phone interview, you will have one big advantage, which is that you can have extensive notes, a crib sheet, answers to questions, and even ChatGPT right there in front of you and they will never know it. Also, on the phone you do not have to worry about what you look like and whether you are having a bad hair day. The downside of the phone is that you cannot see each other, so it can be hard to sense whether you are making a connection with them, and since they cannot see you, your winning smile is not going to do you much good as a way of showing your interest. Because you cannot use body language and visual cues in a phone call, you need to go out of your way to periodically interject verbal cues like "Uh huh," "Sure," and "Got it" during the conversation. These are the verbal equivalents of nodding when you are face-to-face.

Video screening interviews

While you will likely be adept at the mechanics and use of Zoom and Teams for meetings, there are a number of nuances that are especially important for interviews.

1. They can see you, so dress professionally, just as you would if you were there in person. Business casual is a good general rule for industry jobs. For an academic interview, dress as you would if you were attending a professional scientific meeting in this discipline. Definitely no T-shirts.
2. Lose the fancy video background. The best background for a video interview is an actual uncluttered office-like setting, not a sunset over Maui. If you do not have a neat office behind you, and especially if there is anything that will look weird or distracting (your kitchen sink, a cat, an unmade bed, or any bed at all), blur your background or use a tasteful background photo. The downside is that digital imperfections associated with virtual backgrounds can be distracting to the viewer. Unless you have a green screen behind you, the technology still is not good enough to prevent the occasional loss of an ear or part of your head if you move around a lot.
3. Have excellent lighting—be careful that you are not backlit by a bright window causing your face to be hidden in shadow.

4. Raise your camera up so that it is almost eye level—no one wants to be looking up your nostrils.
5. Make eye contact. Adjust your camera, laptop, and the placement of the video window on your screen so that when you are looking at the image of the interviewer their face is as close to the physical camera lens as possible. Periodically look *directly into* the camera lens when answering a question. This is much harder than it sounds, since we are programmed to look either at our image or that of the person we are talking to. Looking at the camera lens for short periods of time gives the interviewer the impression that you are looking at them. Especially during screening interviews, such impressions (subliminal as they may be) are important.
6. The worst thing you can do is to have two monitors with your camera mounted on one while the image of the person you are talking to is on the other. Then what you are likely to be doing is looking at the monitor with their image on it while the camera on the other monitor is pointed at the side of your face. This will make them feel like they are having a conversation with someone who refuses to look at them, which is just horrible.
7. Screening interviews are as much about making an impression on and connecting with people as they are about information gathering.
8. Because the interviewers can see you looking away from the camera to read notes you have on your desk, or someplace on your screen, it will be apparent and distracting. Reading answers from a ChatGPT window during the interview may seem like a lifesaver, but it will be 100% obvious to the other person because they will be able to see your eyes moving while you scan the ChatGPT window. If you want to take notes, for example, if they ask you to get back to them about something, say "I'm just going to jot that down" while you do it, rather than trying to do it surreptitiously.
9. Pay attention to your body language. Smile on occasion, nod when people are talking to you, and use hand gestures to make a point. Adjust the angle of view of your camera or how far away you are from it so that your hands and arms can be seen. Use your hands to make a point just as you would if you were talking to them in person. Positive hand expressions include showing open palms and smooth flowing motions. These things reinforce the sense of engagement and connection with the interviewers.
10. Do not under any circumstances try to record the interview, either using a Zoom add-in or any other AI tool, as this is intrusive and inappropriate. If they wish to do so, that is their prerogative, and you should not object.
11. Expect that you will get no feedback at all during the interview; do not let this demoralize you. They do not want to give candidates false hope. One strategy I use when I am on a video call with several people is to look for the one with the most expressive face and watch them. Seeing someone nod or smile briefly can help you maintain a positive frame of mind.

Video etiquette

- *Get your camera at eye level*
- *Look at the camera lens periodically*
- *Use your hands*
- *Light up your face*

If they ask, "Do you have any questions for us?" this is a golden opportunity not to be missed. We will be covering a long list of "questions for them" later in this chapter. You will have time for only one or two key questions, so choose carefully. Ask questions that show you have done your homework on the person you are meeting with, the group, or the organization. Questions like "What's the salary?" or "How much vacation time do I get?" are definitely not on this list.

Email all of your interviewers after the interview to thank them for their time and for considering you as a candidate and offer to provide any additional information they may need. If you can, thank individual interviewers for specific things they may have said or asked.

ON-SITE INTERVIEWING FOR BEGINNER TO INTERMEDIATE LEVEL ROLES

Once you have aced the screening interview, the next step will likely be an in-person visit that could run anywhere from a half day to two full days depending on the job and the organization. Make sure they send you a schedule for your visit that includes as much detail as possible. Find out who your host will be, where you will be staying, who will pick you up at your hotel, and, most importantly, what their cell phone number is, because unexpected things happen.

Typical interview visit schedule: Academic position

Day 1

- Fly in – get met at airport
- First meeting with Chair
- Cocktails and dinner despite the fact that you want to go to sleep

Day 2

- Breakfast with someone
- Interviews with faculty
- Seminar
- Lunch with faculty, students, or others
- More interviews
- Chalk talk to faculty
- Department tour
- Meet with students
- Meet with HR
- Meet with Dean
- Cocktails and dinner despite the fact that you want to go to sleep

From the minute you speak with or meet the first person from the host organization, you are being interviewed. Be the best version of yourself. But no matter how friendly the people are, be careful about "letting your hair down" and saying "what you really think," especially about other people, including your current adviser or boss, or organizations.

Before you visit, find out the following:

- Who is coordinating your visit and what their cell phone number is
- Where you will be staying
- How you will get from there to the campus or office
- Is someone picking you up, and if so, who are they? (so you can research them and be able to make intelligent small talk during the trip)
- Their cell phone number (so you can report the unexpected flight delay)
- Your interview and visit schedule
- The names of everyone you will meet with (so you can research them)
- When the visit will be over (so you can make travel arrangements)

BE PREPARED FOR THE UNEXPECTED

All academic and some industry interviews include giving a research or "job" talk (covered in detail below), as well as having one-on-one interviews with faculty or other professionals. In an academic setting, these meetings will ideally happen after you have given your talk so the people you meet may remember one or two things you said. But you also may meet with some people before your talk and with others who could not attend it. So be prepared to give a two-minute overview of what you work on and why it is important using highlights from your research statement or talk. Even people who were at your talk may ask you to begin the interview by summarizing the most exciting things you are working on.

Because you will have studied the background of everyone you are meeting with, you will know what level of detail and technical jargon to use for each. Do not describe your research to a nonspecialist in your field in the same way you would to a specialist. Even if they are in a closely related discipline, they will almost certainly not be conversant with the scientific microdomain that you work in and its terminology or acronyms. As in all parts of the interview, prepare for this by giving a three-minute pitch to a friend who is not in your specialty and see what their reaction is.

THE BASIC INTERVIEW QUESTIONS

Your first meeting, with the Chair of the department or search committee in academia or with your hiring manager in industry, will typically happen early in your visit. Below are some common questions that you may get asked. Prepare answers for all of these.

- **Give me a brief outline of your career.**
  Focus on what is most relevant for this specific job in thirty to sixty seconds.
- **What are you passionate about?**
  Are you ready to say that you are "passionate about zeolite catalyzed propane dehydrogenation"? Do not take questions like this literally. What they are trying to get at is what drives or motivates you, so tell them that. Be honest, but remember that your "passion" needs to align with their needs. Something like "I love doing science, and I love the rush I get when I figure something out. My dream for as long as I can remember is to be a professor, run a research lab, and do interesting science," or "I really want to do something that has an impact on people's lives."
- **Tell me about your current job.**
  Accentuate the positive and be brief. Highlight one or two aspects of what you do that align with the job you are applying for.
- **What do you love or hate the most about your current or previous jobs?**
  Tread carefully here. Find something you loved about each position and forget the rest. Just because they asked about negatives does not mean you have to tell them a horror story, so do not take this literally—"They are asking me to tell them something I hate about my current job. That's easy—my adviser is a jerk. So, I'll tell them that." This is obviously a big mistake. Answer the part of the question that you want to answer—what you loved about each job you have had. With any luck, after you answer, they will go on to another question. If they insist that you say something negative about your current or previous positions, you can always make it about a mismatch between the job and your goals and aspirations. Here are two examples:
  - "As a postdoc I wish I had a bit more freedom to follow up on my own ideas, but I understand that I need to keep focused on the goals of my PI's grant. I'm really looking forward to starting my own lab so I can do that."
  - "I love basic research, but in my current job it's hard to see how the things I work on are relevant to helping people. That's one of the reasons I'm attracted to your company. Your work on treatments for epilepsy is really inspiring to me."
- **What accomplishment are you proudest of in your career?**
  Have one ready to talk about. If you are applying for an academic position, it is OK to focus on an approach, project, or idea that was uniquely yours. If it is an industry job, find something that was a team effort in which you made a contribution to the team's success.
- **What's your ideal job?**
  If you start describing your ideal job, chances are it will not exist or will be different from the job you are applying for. As with the "passion" question, the intent of this question is for the interviewer to find out what drives and excites you. What you are answering is "Tell me about what's important to you or what you are looking for in a job." This one is a lot easier and less dangerous to answer than the one about an ideal job because you can answer it by working backward—identify the

things that attract you to the job you are interviewing for and this organization and talk about why those things appeal to you.

- **Where do you want to be in five years?**
  Be aspirational but realistic. If you are early in your career, a safe strategy is to say that you are still figuring this out. This shows that you are not restricting yourself to a narrow career path too early in your journey. If your answer is too specific, i.e., "I see myself starting my own company" or "I'd like to be a tenured professor at Stanford," you can come across as being unrealistic or interested in a path this hiring manager has no desire to help you with. The safest answer to this one is that you see yourself advancing in *this* organization.
- **Why are you looking for a position right now?**
  Instead of "My current job is a dead end with no opportunities for advancement," try "I love my current position but because of the structure of the company it looks like I don't have realistic opportunities to grow there." Instead of "My grant didn't get renewed, and I'll be without funding in six months," try "This is a tough funding environment and like many others I'm looking at many options for the coming year." As with the question about what you dislike about your current job, never say anything disparaging about your current position. They will have no way of judging the truth of what you say, and you could come off as a constitutively disgruntled individual.
- **What do you think are the key skills that you bring to this position? Why are you the best fit?**
  Here is where your research on the organization and the job will pay off. You need to be able to describe what you think they or the position requires and match as much of that as possible to your skills. As noted earlier, do not worry if your skill set is not a 100% match for their needs. Focus on the areas of overlap. If they ask about your abilities in areas where you have minimal or no experience, answer honestly but positively. For example, "I've not taught an entire course myself, but I have been a teaching assistant and am up to date on my evolutionary biology, so I'm certain I'll be able to put together a syllabus for the EVO 101 course very quickly for my third semester here." Or "I think my references will tell you that I'm a quick study. So, I have no hesitation in saying that I will come up to speed very quickly on requirements for working in a sterile environment."
- **Why are you interested specifically in this job/company/position?**
  Know a lot about the organization and describe how *you can help them.*
- **What kind of salary are you looking for?**
  Do not discuss salary during early interviews. It is possible that all they are looking for is an indication that you do not have unrealistically grand expectations. Here are several scenarios.

  For a job where you could find no information on what the salary range might be, your best bet is something like this: "Well, I have not been able to find what the salary range for this position is, and I don't want to guess." Or "I couldn't find anything posted about the salary range, so all I can say is that I would expect a salary consistent with what others with my experience and in similar roles make here." If they press you for a number, keep finding polite ways to decline ("Gosh, I really can't say" or "Can we put that off until I can do more research?"). The danger is that if you quote a very high number, you risk being seen as unrealistic or unaffordable. If you quote a number below what they are willing to pay, then you have done yourself a disservice. So, unless you have a very good idea of the salary range, it is best not to give a number at all.

More and more companies are posting salary ranges when they advertise jobs. In that case, saying something like "Well, I see that the salary range is $80,000–$120,000 a year, and given my background I'd be comfortable in the upper end of that" is all that is needed. But your first interview, whether it is a screening or an on-site interview, is not the time or place to be negotiating your salary or startup package. It would be unusual for you to get asked to do that, but you need to be prepared with deflection tactics just in case. Queue up phrases like "Wow, it's a bit early for me to answer that" or "Can I get back to you on that one?"

Related to this question is "How much do you make now?" In 2025, it is illegal to ask this question of job candidates in twenty-three states, but that does not mean they will not ask. We will cover responses to this question in Chapter 9 on negotiating a job offer, since that is typically where it will come up. For now, "Oh, gee, the job I have now is nothing like this one. I have a couple more questions for you if I may …" And leave it at that.

- **Do you have children?**

  Answer this one at your discretion, along with the related question "What do you do for childcare?" While not technically illegal to ask, some unscrupulous employers use this question to screen out people (especially women) with small children on the grossly mistaken belief that they will require more time off or have less flexible schedules. If you can prove that this is their intent, then the question is illegal and constitutes discrimination. Proving intent, however, is difficult if not impossible and obviously not something you want to be thinking about during an interview.

  As much as you would love to say "It's none of your business," there are better approaches. A simple "yes" with no further detail may send the message that you think this is an irrelevant topic. If they ask for details, then "We have two toddlers." And again, leave it at that. If they get aggressive and ask, "This is a demanding job. What do you plan to do for childcare?" then you can go full tiger parent: "We have always had excellent childcare as well as back-up options. If you are wondering whether having children will impact my commitment to this job, I can tell you categorically that it will not. I will match my commitment and productivity side by side with anyone else in your company and you will say the same after I've been here for a year. It's just not an issue." Depending on the response you get and on what you learn from others who work there, such grilling about kids might incline you to question whether this place is for you.

More responses to inappropriate questions are covered in Karen Kelskey's excellent book *The Professor Is In*.[37]

DOMAIN-SPECIFIC QUESTIONS

Academia

Academics will be looking for numerous characteristics and capabilities they want you to have beyond being a terrific scientist doing fundable research. These include a genuine interest in their institution, a willingness to teach and serve on committees, and whether you will be a good fit personality-wise. Here are some common academia-specific questions.

- **Why are you applying here?**
  You need reasons that show you have done research on them and that explain why you are applying to and are attracted to this specific place.
- **What will your first grant proposal be on? Where would you apply for funds? What will you do if it is rejected?**
  This should flow logically from your research statement and will be highlighted in your chalk talk (both covered below). Get guidance from your current adviser if you are a postdoc.
- **How does your research address the big unanswered questions in your field?**
  As above, this should be explicitly addressed in your research statement and will be a focus of your chalk talk. Here, you will need a thirty-second summary.
- **How will your research be different from your current adviser's?**
  Be prepared for this. We cover it in detail below under "independent thinking/ autonomy."
- **Why are you a good fit for this department?**
  Your research on the department should show you what their strengths are and how you can either complement those or fill some gap. The position description should tell you what or who they are looking for.
- **Are there people in our department with whom you might want to collaborate? Who are they?**
  Here is another place where your research on the department will pay off. Have at least one person in the department who you think could be a possible collaborator but be very hypothetical. It is presumptuous to suggest collaborating with a faculty member with whom you have never discussed it.
- **How much space will you need?**
  This will be subject to negotiation once you get an offer. If you get asked this early in the process, it may signal that they are trying to determine whether they can accommodate your needs. The problem with tossing out a number before you are ready is that if you are too low, you could get locked into a space that does not meet your needs, and if you are too high, they could conclude that they cannot accommodate you. Your best bet in early-stage discussions is to say something like "I'm still working that out."

- **Tell me about your teaching experience. Have you ever had to create a course or teaching unit on your own? Tell me about that.**
  Covered below.
- **What is your teaching philosophy?**
  Covered below.
- **Which of our existing courses can you teach?**
  You need to have read their course catalogue and have one or two you could teach when the time comes (covered more in Chapter 5).
- **What new courses could you develop?**
  Have a suggestion based on your research of what they already teach and what you could add in your specialty. If they do make you an offer, part of your negotiations will be that you are not required to develop any new courses for your first one or two years (covered further in Chapter 9).
- **What experience do you have mentoring students? Can you give me an example of a mentoring experience you have had?**
  Covered below.
- **How will you engage undergraduates and graduate students in your research program? What kinds of projects would they work on?**
  Have a few projects in mind that require varying degrees of expertise suitable for both undergraduate and graduate students.
- **Have you ever chaired a committee? Have you ever been on a committee? Tell me about it. What did you contribute/accomplish?**
  This is another "look for the intent" question. All academic jobs will require you be on one or more committees (admissions committee, search committee, curriculum committee). If you cannot provide examples of having been on actual committees, find something you did—being a member of a club or organization—and come up with a way to convey your willingness and interest in being a "good citizen."
- **Do you need specialized equipment?**
  This is another question that, if asked early in the interview process, could be their way of making sure they can afford you. This will all come out when you submit your startup package details, but if you need a $5,000,000 NMR machine, they will want to know sooner rather than later.
- **Have you been involved in diversity, equity, or inclusion efforts? Tell me about that.**
  The last was an element of many academic job interviews until 2025, when the US government began to pressure educational and other institutions to eliminate it. Many or most have done so. But if we have learned anything about the nature of US politics and its impact on institutional polices, it is that things change. Just because this line of questioning is disappearing now, it does not mean that by the time you send in your application it will not have changed back. Better at least to have thought about this issue and have an answer at the ready.

Industry

General questions

Some common questions to prepare for in an industry interview include the following:

- **Why have you chosen to apply to *this* company?**
  You must be able to talk about what you know about the company and what appeals to you about what they do based on your research.
- **Why do you want to move into industry?**
  Tell them what attracts you about an industry job—the opportunity to contribute to solving a real-world problem, working as part of a team as opposed to a one-person project, or the opportunity to advance in your job are all good places to start.
- **What would you do if your boss made a decision that you strongly disagreed with?**
  Mention being aware that differences of opinion are common in science, that you respect others' views, and that you know how to navigate disagreements in a professional matter. Let them know you understand that when a consensus is lacking someone has to make a decision to move projects forward. In industry, such decisions may be made for reasons that have little or nothing to do with science. Some scientists find this frustrating, especially when they do not understand or are unaware of the reasons. Saying "I understand that sometimes decisions get made based on changes in strategy or company focus" is all that is needed.
- **Tell me about your management experience.**
  Companies frequently need to fill managerial or supervisory gaps quickly, so hiring staff who have good interpersonal skills or who have done some supervision is a plus. If you have such experience, even having managed a student or technician, highlight that. Chapter 12 covers more questions and answers regarding this theme.

Technical skills

In academia, your technical skills do not get probed per se unless you are applying for a technician or postdoc position. For faculty-level jobs, your scientific and technical bona fides come out through your publications, your job talk, and your one-on-one interviews. No one is going to ask you to describe how to set up a PCR reaction unless you are applying for the job of research assistant or for a specialized role in industry. This is why it is critical that you have been 100% honest in your résumé and cover letter. The last thing you want is for the interviewer to ask some geeky technical question after you have claimed to know how to perfuse a neonatal mouse or calibrate a Cherenkov counter when all you did was watch someone else do it.

If you are applying for an entry-level job, you should *expect* questions probing what you know. Every interviewer has their favorite way of going about this, including how you would troubleshoot some technical mishap, like a PCR reaction that did not yield any products; how you would go about solving a coding problem; how you would design a simple circuit; how you would explain what a stress strain curve is; and pretty much anything else. There is no way to prepare for these other than to know what you say you know. If it happens that you cannot answer a question, you can always brush it off to nervousness: "I'm sorry, I guess I'm feeling a bit nervous right now and I'm blanking on the answer." If you are lucky and they are empathic, they will move on.

What if you do not have the technical skill they need? We encountered this in the previous chapter where the position description called for GLP experience, which the applicant lacked. We advised not calling this out in the application letter. However, what should you do if asked during an interview? If the item was an absolute deal-breaker, you likely would have been eliminated early on. But even in cases where the item is specified as a must-have, people change their minds when faced with an otherwise stellar candidate. One way to handle this is as follows:

I have a general familiarity with GLP lab work (and seeing this in the position description, you would have gained such familiarity through research before the interview). *I can tell you that I'm a fast learner with respect to lab procedures and protocols and I am 100% confident that I will be up to speed on this before you know it. I was cognizant of this requirement in the posting, and I can sincerely tell you that I wouldn't have applied for the position if I thought it would be an issue. It won't be.*

Last year our group was selected to be the first to adopt a new electronic lab notebook format. I was given the responsibility of getting trained by the provider and then rolling it out to the analytical group, which I did in pretty close to record time. I think I'm pretty good at stuff like that. I'm sure with the GLP thing that I'll pick it up quickly.

Skill tests

Beyond technical questions already discussed, some candidates report being asked to actually perform a task during an interview. A cursory search of Reddit suggests that in certain disciplines, most specifically coding, skill tests are sometimes used, as in being asked to debug a program or outline a strategy for an algorithmic approach to a technical problem. There is little or no hard data on the frequency of such questions, but they are most commonly asked of junior or entry-level personnel. This can be stressful, but if a company is hiring you to perform a task that requires specific skills, it makes sense that they will want to verify those skills before they make an offer. As noted in Chapter 1, this practice will become more prevalent as applications and letters crafted by AI become the norm, requiring employers to resort to in-person confirmation of skills.

As the use of AI becomes embedded in nearly every aspect of work, it stands to reason that the ability to use AI optimally will itself be a valued skill. A recent report indicated that Facebook's parent company Meta is experimenting with allowing applicants for coding jobs to use AI coding assistants during their interviews.[38]

MEET THE TEAM

Whether you are applying for a role as a bench scientist, faculty member, or senior science manager, you will be working for and with others and some will be working for you. Nearly all employers will give you and their staff the chance to meet either in formal interview settings or informal situations like lunch or dinner. Beyond having a wide array of people opine on your scientific qualifications, these occasions are also about your "fit" with the organization or department. This will mean different things to different people, but at the very least they are asking what it will be like to work with you, hang out with you socially, be on a project or committee with you, and sit next to you for eight hours on a transatlantic flight.

We have already advised you to make a one-page profile of everyone you will be meeting with, so you have something to say or a question to ask them. Even if the question is superficial, the fact that you have researched them will show that you are a serious candidate. If you are going to dinner with members of the search committee, find out who will be there and be prepared to ask them questions that show a genuine interest in them and their job.

As indicated in the "your questions for them" section, it is OK to ask questions like "What's it like to work here?" or "What's it like to work for [insert name]?" If in response you get silence, change the subject. If you get a vituperative tirade, start by assuming that you are talking to one disgruntled employee. If others have the same reaction, proceed with caution if they offer you a job.

Be careful about letting your hair down, especially if asked about your current job, adviser, or organization. Negativity never plays well and can be especially damaging during an interview. This is yet another reason for you to limit your alcohol consumption if they take you out to dinner. Your hosts may cut you some slack if you make an ill-advised comment after a couple of cocktails, but do not count on it.

If you are unsure of how to comport yourself in social situations, study up on the basic rules of American etiquette, especially if you are new to American culture.[39] When all else fails, being a bit more formal than the situation requires is invariably better received than being inappropriately casual. Pay attention to matters of dress and personal hygiene. People are going to spend a lot of time in your presence during your visit and will notice small matters relating to grooming, dress, and cleanliness, either consciously or unconsciously. If you are not in the habit of wearing antiperspirant, now is the time to start. You will be nervous, but there is no need to broadcast that via sensory data. Other small things, such as having a modest shine on your shoes, freshly pressed apparel, and a recent haircut, are also worth attention.

INTERVIEW QUESTIONS TO PUT OFF FOR A LATER TIME

Until they decide to make you an offer, you have little or no leverage. Bringing up contentious issues too early could make them think that hiring you is going to be a pain in the neck. The time to do that is once they have made you an offer. Then they are committed, and they really do not want to have to start from scratch with another candidate, or worse, a new search. Things to put off till later include the following:

- Salary and benefits
- Your specific needs for lab space and fancy or high-cost equipment
- Special needs for spouse, i.e., the "two-body problem"
- Your desire to put off starting the job for some period of time or to work remotely for part of the time

BUT DO THEY LIKE YOU?

Some interviewers' evaluations will be made not on the basis of your qualifications but on "gut feelings," mental heuristics, and biases, often including that they did not feel a "connection" with you.[40,41] While you cannot counteract such biases, you *can* work to create a personal connection with the people you meet. Make eye contact and smile during meetings and dinner and display visible signs of understanding by your body language (facing them square on and nodding) and by verbal cues ("Uh huh," "Got it," "I understand," etc.). That is, treat the whole process like a networking event! Even if you do not get the job, you might make valuable contacts and possibly even friends.

CHAPTER 6 SUMMARY

Do extensive research on every person you meet

- Create individual profiles with photos, backgrounds, and research interests, and prepare specific questions for each interviewer to demonstrate you have done your homework and can make meaningful connections.

Treat screening interviews as seriously as full interviews

- Do not wing twenty- to thirty-minute phone/video screenings since they determine if you advance. Prepare fully, pay attention to the technical setup (lighting, camera angle, background), and have one or two thoughtful questions ready.

Prepare polished answers for standard interview questions

- Have ready responses for your career overview, what you are passionate about, current job description, accomplishments, five-year goals, why you are job searching, key skills, and salary discussions (deflect salary talk until you get an offer).

Understand domain-specific expectations

- Academia focuses on your research program, teaching philosophy, grant writing, collaboration potential, and departmental citizenship. Industry emphasizes technical skills, management experience, team fit, and ability to work within corporate constraints.

Maintain professional behavior throughout the entire visit

- You are being evaluated from the moment you arrive until you leave, including during meals and casual conversations. Avoid negativity about current positions, limit alcohol, and demonstrate genuine interest in everyone you meet.

Save sensitive topics for after receiving an offer

- Do not discuss salary, specific lab space needs, expensive equipment requirements, a spouse's job needs, or start date flexibility during early interviews when you have no leverage.

The "fit" factor

- Hiring decisions are often made subconsciously. You can improve your chances by creating genuine connections through preparation, eye contact, active listening, and positive body language.

7 A Deeper Dive

Acing Behavioral Interview Questions

Advanced interview questions. Them for you and you for them.

- **Managerial skills**
- **Behavioral interview questions**
  - Responsiveness to feedback
  - Ability to manage emotions
  - When you get stuck
  - Are you a team player?
  - Conflict management
  - Independent thinking/autonomy
  - Miscellaneous oddball questions
- **Your questions for them**
- **Chapter 7 summary**
- **Appendix 1. Your questions for them: Academia and industry**

There is no predicting the kinds of questions you may get asked in an interview, so it is best to prepare for anything that might come your way. The following sections cover topics you should prepare for that go beyond the basic interview questions in Chapter 6. In addition, this chapter includes an appendix that contains extensive lists of questions that *you* should be prepared to ask your interviewers in both academic and industry settings.

MANAGERIAL SKILLS

You may get asked questions about your ability to manage or supervise others, especially in industry. Even if the job you are applying for is not specifically a managerial one, it could include managing a technician or lab assistant. If you show competence in this, you could become a candidate for promotion to a supervisory role. So, pre-vetting people for these skills means that the organization has an existing pool of talent ready to move up.

If you are lucky enough to have either managerial or mentoring experience, you need to be able to speak about these fluently and convincingly. If you have little or no prior management experience, hark back to what we said earlier and find something you have done in your life that put you in a managerial, supervisory, or mentoring role, and talk about that.

Organizations are taking a risk when they put you in charge of employees, and the best of them will go to great lengths to make sure that you are not a managerial train wreck waiting to happen. For a primer on what it means to be a good manager, read *Lab Dynamics*, Chapters 5, 6, and 11. These chapters will give you the vocabulary you need in order to talk about the qualities you value in your management style.

For reasons that are spelled out in *Lab Dynamics*, you will almost never get asked this for an academic position. Most people who will be interviewing you, including faculty, do not think a lot about the process of managing, and many would have difficulty articulating what good management even looks like. This is unfortunate because it leads to academic scientific organizations populated by staff who excel at science and either fail at or struggle with managerial or interpersonal matters. Nonetheless, you should prepare for any management-oriented questions you may get asked, most likely from someone in human resources.

1. **Tell me about your management style.** They are asking you for buzzwords and bullet points, and that is what you should give them until they get more specific. Do a bit of reading about the basics of managing scientists and learn to use words and phrases like "employee engagement," "goal setting," "useful feedback," "holding people accountable," and "empathic leadership."
2. **Tell me about a time you managed someone who was not performing up to your expectations. What did you do?** You may never have had such a situation, but your answer is not going to be "That never happened to me." It is going to be "I've been lucky enough not to have experienced that firsthand, but I have some ideas how I would handle it based on experiences I have had."

   Then start talking about how you would never let the situation get to this point in the first place because you would be meeting with the employee regularly, and giving them feedback and help with their performance, before it got to the point of being a liability. Add to that "Of course, it's always possible that someone just is not in the right role and no matter how much help and guidance you give them they are not going to be performing up to your expectations. I believe that those are rare situations, but when they do happen it's best to part ways sooner than later. Chances are if I'm not happy with their performance, they are not happy with the job."

We will cover this topic in more depth in Chapter 12.

BEHAVIORAL INTERVIEW QUESTIONS

When you get interviewed by a recruiter, a specialist from human resources, or a sophisticated hiring manager, you will likely get one or more questions designed to probe some aspects of your personality. As we discuss later in Chapter 12 on interviewing for senior executive roles, such questions are designed to give the interviewer a sense of what makes you tick and how you might behave in difficult or stressful situations. Suzanne and I discuss asking such questions in *Lab Dynamics*, where we give examples of questions as well as what to look for in the answers that candidates give. Here the same issue is covered from the candidate's perspective.

I often get the following question in my workshops on how to interview and hire scientists. "Carl, you are presenting all these behavioral questions and telling us what to look for in the candidate's answer. What's to stop a candidate from reading *Lab Dynamics* and then using what they read during the interview? Aren't they just gaming the system by pretending to be someone they are not?" The answer is no, that's not gaming the system, it is studying and learning. If a candidate has gone to all the trouble of reading *Lab Dynamics*, of studying and remembering what kinds of behaviors work best in a variety of difficult situations and how to speak about these behaviors, they have become authentically knowledgeable, and you should hire them. This shows a level of commitment that will put them in the top 10% of applicants.

The questions listed below are called behavioral or situational interview questions because they seek to identify and select for certain valued responses or behaviors in specific situations encountered at work. Do not view the suggested answers to these questions as an easy way to get through an interview without understanding or buying into what the answers mean. The guidance and answers are meant to enable you to develop and grow as a person so that you can function at your best in complex scientific environments.

Responsiveness to feedback

This trait is one that savvy hiring managers try to identify and select for in candidates. They understand that no one knows everything and that no one gets everything right all the time. Good managers accept this and use thoughtful feedback to help employees see what they can do better and guide them in how to do that.

Interviewer: "Have you ever been told that your work wasn't up to standards or that you needed to do better? What was the situation? How did you handle it?" and "Tell me about a time when you made an actual mistake. What happened?"

Your goal is to demonstrate that you understand that at times everyone does things that require feedback from their manager about what they did and how they can do better. If you accept such feedback with grace and humility, good for you. If you do not, now is the time to start.

You: Be ready with a story about a time when you got feedback, be able to articulate that you are eager to hear feedback, and show that you know how important it is even if it is delivered punitively (as it sometimes is by unskilled managers) and is hard to hear. Demonstrate this skill when you respond to questions during your job talk or interview. Are you open to probing questions? As in "Yes, that's definitely an alternative explanation. Thanks for the suggestion." Or are you defensive, as in "No, we thought of that. Not likely. As I was saying ..."

Ability to manage emotions

Emotional dysregulation, including outbursts, petulance, sulking, and passive-aggressive behavior are behavioral characteristics that can hamper your career. Being able to identify and manage strong feelings before they overtake you and become dysfunctional behaviors is a key element of being emotionally intelligent. We all have feelings in response to work-related circumstances. What sets the most valued employees apart is that they can manage these feelings and their behavior.

Interviewer: "Tell me about a time when you got angry or upset about something at work. What triggered that, and what did you do?"

You: Have an example ready of a time when you managed feelings of anger or frustration without "acting out," losing your temper, or blaming someone else. There is no shame in feeling angry or frustrated in response to someone or some situation, as long as you did not let those feelings manifest themselves in angry behavior. Show your ability to self-reflect. ("I debriefed myself after that incident and thought a lot about what I should have done.") During the interview process, monitor your reactions to interview schedule changes, delays, unexpected occurrences, etc. Your goal is to convey that you have high emotional intelligence and stability.

When you get stuck

What do you do when you get stuck? Many scientists value their ability to drill down into problems and figure things out for themselves no matter how long it takes. They take pride in their problem-solving skills. This trait can be a red flag when applying for a job in industry, where finding a path forward quickly is more valued than finding the best or most beautiful path forward. Industry values pragmatists more than perfectionists, those who seek help in finding a way forward more than those whose pride demands that they figure it out for themselves.

Interviewer: "Tell me about a time when you were stuck on a difficult problem. How did you resolve it?"

You: Here is where a bit of honesty can work in your favor. If you are one of those scientists who love to hunker down and dig in no matter how long it takes, talk about how you know this about yourself, and that you understand that in an industry environment that is not the best approach. Show that you understand the need to find the shortest path to a solution and that you are committed to that.

Are you a team player?

This question or a variant of it will often come up in industry interviews, and for obvious reasons. In industry you will inevitably be part of one or more teams, and you will need to be able to function well in that setting. Interviewers recognize that scientists just coming out of an academic environment may not be used to working in a team setting. If you were part of a project team in academia, then you are one step ahead of the game. If you were not, you need to familiarize yourself with what to expect when you start to work on a project that is not yours, one in which your individual contributions may be obscured by those of the team as a whole. If you are OK with this, you need to be able to talk about it convincingly.

Interviewer: "Tell me about a time when you worked as part of a team on a project. Was that difficult or enjoyable?" And "Do you like working on a team or do you prefer your own projects?"

You: It is OK to convey a balanced approach—you enjoy the sense of independence and challenge of working on your own, but you also have had positive experiences as a team member. Have ready a story about a satisfying team project you were part of and why you liked it.

Conflict management

Even if you are a newcomer to science, you have probably experienced a situation where there was some disagreement over how to interpret data or what the next experiment should be. Experienced scientists understand that such conflicts are natural and inevitable. If that is not you, then get to work reading Chapter 3 of *Lab Dynamics*. Learn to use the language of negotiation when you speak about such differences. Ideally, you need to believe and convey your understanding that disagreements are natural and that you work hard to understand the other side's viewpoint.

Interviewer: "Can you think of a time when you had a disagreement with someone in the lab? Tell me how you handled it." Or "Have you ever disagreed with a decision that your supervisor or organization made? Tell me how you managed that disagreement."

You: Have a story ready that illustrates your ability to negotiate solutions, to see the other person's perspective, and to manage the conflict in order to get the work done. Be alert to a tendency to blame others or assign fault.

Independent thinking/autonomy

This line of questioning will come up most commonly in industry interviews for intermediate to senior positions. It may also come up in academia where they are probing whether your graduate or postdoctoral work was based on your mentor's ideas or whether you played a role in setting the direction of your research. Even if you have not had much opportunity to demonstrate independence, you should be able to think of a situation where you had an idea for an experiment, or for an interpretation of one, that led in an interesting direction.

Interviewer: "Did you ever have a project for which you chose the focus or strategy?" Or "Have you ever been at a point in a project where you had to make a decision about changing direction or strategy? Tell me about that."

You: Have a story that shows you can come up with ideas or ask insightful questions that lead in new directions. Ideally, it should show that you can come to independent conclusions and that you are OK with taking risks and thinking independently, even while being a team player and seeking broad input.

For an academic position where you are expected to establish an independent research program, they might probe your independence from your adviser or mentor.

> **Interviewer:** "What distinguishes you and your future research program from that of your current adviser?"

You need to have an answer that shows exactly how what you plan to do is different from what your adviser is doing and is not just a trivial extension of your postdoc work. Unless you can do that, the search committee may worry that if that research direction plays out, you will end up at a dead end. Ideally, your research program needs to have new directions coming from ideas of your own.

There is nothing wrong with continuing with a productive research area that you started as a postdoc, especially if it is not going to be a major focus of your adviser's lab that they plan to continue with. Unfortunately, some advisers have a propensity to appropriate their postdoc's ideas, especially if it is an exciting new direction. Their reasoning is that this was a collaboration, it was in their lab and they paid for it, so they have rights to the project. In such cases, you will need to negotiate a creative way to split the project into well-delineated foci so that both of you can continue on your own paths. More enlightened advisers will allow you to start a project that you both agree up front you can take with you.

Miscellaneous oddball questions

People still get asked all sorts of "trick" questions by interviewers who think they are being clever or sophisticated and who read questionable psychological meaning into the answers. Here are a few:

- **"What's your biggest weakness?"** This question is the hallmark of a lazy interviewer, but you should not betray that you know that. A Google search will instantly tell you that the answer to this question is to name a weakness that is actually a strength, i.e., "Well, I tend to be very highly goal oriented when working on a project. I'm single-mindedly focused on success, and when I'm like that I know I can be very demanding of my staff. I'm aware of this tendency and I'm really working on balancing the company's need for progress against employee burnout." I'm not suggesting you use *this* answer because even a moderately sophisticated interviewer will see through it, but you get the idea.

  As we have with previous questions, ask yourself what the interviewer is really looking for. They are clearly not expecting you to reveal some fatal professional or personal deficit that puts you out of the running for this job. ("Honestly? I never write anything in my lab notebook.") More likely, they are looking for a sign that you know we all have something we could do better—that you have some self-knowledge and humility. Knowing one or two of these things about yourself, and acknowledging that you are working on them, is a strength. Do a little self-reflection and find something that you could be better at and that you could work on. Identifying something that you need to be better at is not a weakness, it is a learning opportunity. For example:

"I'm not sure I see this as a weakness, but more as something I'm working on. I'd really like to be better at (… Python; reading financial balance sheets; being a better mentor …). This hasn't affected my work up to now, and I want to make sure that doesn't in the future" could do it.

- **"What would your current adviser (or colleagues) say about you if I were to call them on the phone?"** This is another question that you should not take literally, e.g., "Well, I've never really taken a poll, so I can't say with any degree of precision what they think of me." Use your everyday experience of how people interact with and work with you and extrapolate: "I think people respect my scientific skills and knowledge and trust my judgment because they often ask my opinion about tough technical matters. I think they would say I'm a good colleague and collaborator."
- **"I think you know one of the other people we're considering for this job. What do you think of them?"** Never say anything bad about another candidate. Period.
- **"Now that you've spent the day with us, what are your reservations about coming here?"** Even if you do have major reservations there is absolutely nothing to be gained by bringing them up. The time to bring up such things is when they offer you the job when you can use your concerns to negotiate a better offer, provided they are not deal killers. Try instead, "I have to say I can't think of any right now. This place looks terrific, and everyone has been very hospitable."
- **"Do you have other job offers? Where are they from? How would you compare those places to us?"** There is nothing wrong with letting them know that you are looking at other opportunities but there is also nothing to be gained by revealing anything else. "Well, as you might imagine I'm looking at other places, but at this stage it's too early to say how those will turn out." If they press you further: "You must be looking at that position at Stanford; it's practically got your name written in the job description," all that's needed is "Yeah, I definitely saw that." Then stop talking or change the subject. The principal reason not to discuss other places you are applying to is that if you get eliminated from the running for those jobs or the job gets offered to someone else, you are at a strategic and psychological disadvantage if these people find out that your options have been reduced.

YOUR QUESTIONS FOR THEM

At some point during your interview, you are going to get asked "What questions do you have for me?" If your answer to this question is "None, I'm good," you may as well pack your bags and go home. You need to prepare questions for this interviewer and everyone else you will be meeting that day. This shows that you are a serious candidate and that you have studied the job, the organization, and its people. There will be different questions to ask of different people and for different jobs.

For both academic and industry positions there are literally dozens of questions you could ask so you can better understand the job. The appendix to this chapter lists a number of them. There are more there than you can or should ask of any one person in a given meeting. Nonetheless, at some point you need to get all or most of these answered. Some can wait until they make you an offer, but some you will want answers to early on. The questions are divided by sector (academia and industry) and further organized by the position or role of the person you are talking to. Pick a few from each category that are most relevant to you, the institution, and the job you are applying for.

CHAPTER 7 SUMMARY

Prepare for management-related questions even for nonmanagement roles

- Companies often ask about supervisory experience, management style, and handling underperforming employees because they want to identify promotion candidates and ensure you can supervise technicians or assistants.

Master behavioral/situational interview questions that reveal personality traits

- Be ready to discuss specific examples of how you handle feedback, manage emotions under stress, work in teams, resolve conflicts, think independently, and overcome obstacles—these questions assess emotional intelligence and workplace behavior.

Handle oddball and trick questions professionally without taking them literally

- Questions like "What's your biggest weakness?" or "What would colleagues say about you?" are looking for self-awareness and humility, not literal confessions of flaws or speculation about others' opinions.

Always have thoughtful questions prepared for every interviewer

- Saying you have no questions is a major red flag. Prepare different questions for different roles (department chairs, faculty, hiring managers, peers) that demonstrate you have researched the organization and are seriously considering the position.

Tailor your questions to academia versus industry contexts

- Academic questions focus on tenure processes, teaching loads, research support, and departmental culture, while industry questions emphasize project scope, team dynamics, company finances, and growth opportunities.

Use questions strategically to assess fit and gather negotiation information

- Your questions should help you understand the work environment, identify potential red flags, and gather information you will need later for salary and benefit negotiations once you receive an offer.

APPENDIX 1

Your questions for them: Academia and industry

There are far more questions here than you could ask any one individual. Despite this, you really would like to know the answers to most of them. Some will get answered without being asked during conversations over the course of your interviews and site visits. For the ones that remain unanswered, select those that are most important to you.

Academia

Questions for the department chair, chair of the search committee, or your faculty host

1. What is your vision for the department/school? Where is it going in the next five or ten years?
2. If you were coming to this department, what would you want to know?
3. What sorts of programs are available for new faculty members? Is there a formal mentoring program? Orientation? Written faculty handbook?
4. Is the department growing or shrinking? Which subfields?
5. Beyond my startup package, is there department research support available and, if so, what kinds?
6. Are there ongoing tensions regarding social or political matters that I should know about?
7. Is there financial support for summer research by undergraduates?
8. Are competitive, within-institution seed grants available beyond my startup?
9. Regarding undergraduates and graduate students, who applies? Who is accepted? Who comes here? Where do they tend to go after graduation?
10. What is the general teaching expectation over time? How many courses per semester?
11. What would you like the candidate for this position to teach over the next three years?
12. What are the expectations for the summer? On campus? Is field work OK?
13. What are your plans for future hires in this department?
14. How often does the department meet to discuss departmental business?
15. How are departmental decisions made?
16. How is chairmanship determined? How long is the term?
17. What nonteaching expectations are there?

Questions for faculty members

1. What do you like best about this place?
2. How do you think this place could improve?
3. How would you describe the culture here?
4. Are there currently ongoing tensions regarding social or political matters that I should know about?
5. What are you looking for in this new hire? Is there unanimity in the department or search committee about this?
6. What does a newcomer here need to know?
7. I read about your fascinating research. I'd love to hear more!
8. What do you teach?
9. How much flexibility is there in what an individual teaches?

Questions about promotion and tenure (for chair and faculty)

1. What is the schedule and mechanism of promotion review?
2. How much of the department is already tenured? Is there a quota?
3. Who decides tenure and how?
4. What kind of guidance is provided about the tenure and promotion process?
5. On what criteria are tenure decisions made?
6. Are the criteria written down and distributed to new faculty?
7. What are the unwritten criteria for promotion?
8. What percent of junior faculty succeed at getting promoted?
9. When do people come up for tenure?
10. What are the criteria for promotion to full professor?
11. How is faculty performance assessed each year? Does this relate to the evaluations for reappointment and promotion? Are there annual reviews before tenure?
12. How is teaching quality evaluated? What kinds of evidence are considered?
13. How are interdisciplinary efforts assessed? What about activities that cut across the research, teaching, and service missions?
14. What separates a good faculty member from a great one at this institution? What kinds of attitudes or work habits make people successful here?

Questions about teaching (for faculty, chair, or dean)

1. What administrative help would be available for teaching-related matters? For research, grants, publication-related matters?

2. How much flexibility is there when courses are scheduled (within a week, within a year)?
3. How big are classes? Are teaching loads weighted by the number of students in the classes?
4. How much time will I have to prepare my first course? What kind of teaching relief do you provide for new hires?
5. Can I teach a seminar instead of an actual course for my first teaching assignment?
6. Are TAs available? For what courses?
7. How are TAs trained?
8. How much funding is there for courses, particularly new ones?
9. How do ideas for new courses get processed?
10. What are the teaching labs like?
11. Are there computer teaching labs? Software? Support personnel?
12. What are typical teaching opportunities across the college?
13. How much variation is there across departments?
14. Are teaching training opportunities available? What kinds? How often?
15. Can one buy out of teaching obligations?
16. How does undergraduate advising work?
17. How does graduate advising work?

Questions about students (for faculty, chair, or dean)

1. What kind of graduate students do you attract?
2. Where do they go once they finish their degrees?
3. What support is available for graduate student research? Who pays for graduate students? Are there any training grants?
4. What support is available for undergrad research?
5. Do students have a say in faculty search decisions?
6. Do students have a say in department administration?
7. What interdisciplinary collaborations are already in place? (Are there any training grants, for example?)
8. How many undergraduate majors come through the program each year?
9. Are students motivated? What do they do after graduation?
10. Are work study students available?
11. Is there student travel support? How often? How much?
12. Do all majors do research? Do they have to submit a thesis?

13. Are there any curriculum changes in the works?
14. Do grad students who TA have time to get their own research done?
15. How are most grad students funded?

Other important details

1. Can I see the available research lab space?
2. How much office space and lab space are there for graduate students?
3. How long will it take to remodel my lab? Who pays for that?
4. As I prepare my detailed research budget, what additional institutional or departmental costs should I know about?
5. When do you expect to make a decision on this position?

Industry positions

Questions for the hiring manager

1. Is this a new position or a replacement? If it's a replacement, why is the previous person gone? If it's new, why was it created?
2. Who will I be reporting to?
3. Who are the team members I'd be working with? Will I get to meet them?
4. Can you give me some idea about the projects I might be working on? Are they ongoing or are you starting a new effort in some area?
5. The position description is pretty broad—what do you consider to be the key skills needed to succeed in this role?
6. If they have given you a schedule of who you will be meeting, run through it with them to make sure you understand who everyone on the list is.
7. What do you like most about working for this company?
8. Is there potential for growth in this role?
9. Is this position funded internally or is there support from a corporate partnership? How long is that expected to last?
10. If I were to take this position, what would you be looking for at the end of my first year that would indicate I've been successful?

For an entry-level position

1. What would a typical day look like for me in this role?
2. Is there any training program for new hires? What does it look like?

For intermediate to senior positions

1. Can you talk me through your major corporate objectives for the coming year?
2. Tell me about your finances. How well capitalized are you? How long until your next fundraising? Will that be dependent on reaching key milestones? What are they?

Questions for peers or coworkers

Most companies will give you an opportunity to meet with potential peers, coworkers, or teammates as part of the interview process. In fact, if they do not, this could be a red flag causing you to question whether this will be the right place for you. A company's failure to do so may suggest a culture in which the input and views of employees are not highly valued.

1. What's it like to work here? Do you like working here? How would you describe the culture? What's the best thing about working here? What could be improved?
2. How do you rate science here? Are they working on cutting-edge problems with the most advanced tools? (It is OK to get specific here.)
3. What kind of flexibility is there in working hours?
4. How would you describe the management style of your supervisor?
5. What's the CEO like? Are they approachable?
6. Do you feel well informed about the direction of projects or about the company as a whole?
7. Is there a lot of turnover here? If so, why?
8. How does the job I'm applying for fit into the team? Is there a specific need that needs to be filled?
9. Are you happy with your compensation?
10. Do you have annual reviews and is there a promotion and bonus structure in place?

8 Showtime

The Three Performances That Can Make or Break Your Career

Research talks, chalk talks, and teaching demonstrations

For an academic position you are going to be asked to perform at least once and more likely three times during your on-site visit. These performances come in the form of a research or "job talk," a chalk talk, and a teaching demonstration. Since being a successful academic involves having a strong research program (research seminar), being able to communicate its value and garner funding (chalk talk), and mentoring and teaching (teaching demo), these performances let the search committee and others see you in action in each of these realms.

By contrast, for a job in industry, a job talk will typically be required only for intermediate-level scientist jobs where they are looking for specific technical skills. Applying for entry-level or higher-level managerial positions in industry will likely shift you back into the no-talk realm. For entry-level industry jobs, it is unlikely you will have material to present, and they are not terribly interested in your presentation skills. For more senior positions, they are hiring you for a much broader set of skills (management, strategy) than would be highlighted in a job talk. Also, by the time you get to the point where you are applying for senior industry jobs, it is not likely that you will have much original research to talk about, and even if you did it would be privileged information owned by your previous employer, and you will not be able to talk about it.

As shown below, in each of the academic performances, what you say and do needs to be consistent with what you wrote in your research and teaching statements. If you mentioned that a key goal of your research is to advance our understanding of exoplanet atmospheres, then you need to talk about that in your seminar and list it as an aim in your chalk talk. If you extolled your ability to use active learning classroom experiences in your teaching statement, you need to actually use that technique in your teaching demonstration or even in your research seminar if there is not going to be a separate

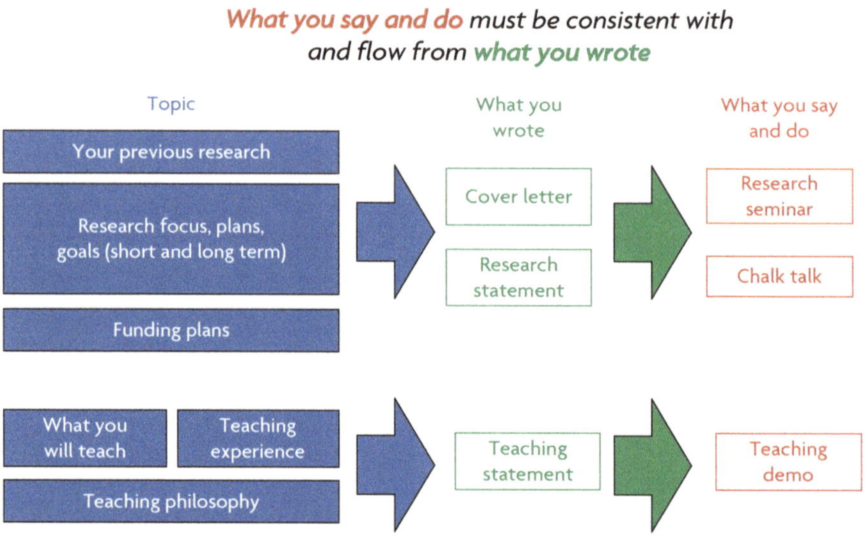

teaching demo. And if you highlighted your interest in collaborating with departmental faculty in your cover letter, you need to have researched those faculty and be able to talk knowledgeably about their work during your face-to-face interviews. Start thinking of yourself as offering a package of integrated skills that together defines you as a compelling candidate. If you have your most compelling features in mind, you will be able to talk about yourself in a way that will make you seem to be more than the sum of the parts.

RESEARCH SEMINAR OR "JOB TALK"

It is imperative that you understand the differences between job talks and research seminars. Too many applicants make the mistake of giving their standard research seminar and end up undermining their candidacy. The job talk is an opportunity for those who will have a say in whether you get an offer to see and hear you in action. This includes faculty and students in academia and people from nonscientific disciplines in industry, all of whom may be asked for their opinion of you. Consequently, you must calibrate your talk to the various levels of expertise in the audience. The table below shows the principal differences between standard research talks and job talks in academia and industry.

Type of talk	Standard research talk	Academic job talk	Industry job talk
Purpose	Focused on research results or methodologies. The primary goal is to disseminate knowledge, receive peer feedback, and engage in academic discourse.	The goal is to showcase the candidate's research program and communication skills.	The goal is to show that you have the technical skills to perform the job.

(Continued)

Type of talk	Standard research talk	Academic job talk	Industry job talk
Audience	Specialists or near-specialists in the field.	Search committee, faculty, and students, most of whom will not be specialists in your field.	Hiring manager, team members, and nonscience staff and executives.
Content and accessibility	Most of the scientific content, terminology, and acronyms will be understandable only to specialists in this subfield.	Focused on your past accomplishments and future research program. Most of the talk must be accessible to a majority present, no matter what their level of expertise.	Typically focused on a project you worked on in your last position. Seek their guidance on the target audience so you can adapt your content.
Things they are looking for in *you*	It's about your research, not you.	A lot. Your research, your delivery, and how you answered questions will all be evaluated. Will you be a good teacher? Did you convey a vision for your future research? Are you an engaging speaker?	Technical expertise in their area of interest; communication skills; responsiveness to questions; and ability to think on your feet.

Do you want to tell them what you did, or do you want to get them to hire you?

Scientists often feel skittish when asked to be even a little self-promoting. We are acculturated to believe that people who speak grandly may be selling snake oil and that serious scientists must speak humbly about themselves and their work. While these suspicions are mostly true for standard research seminars, they are off the mark for a job talk. There you are selling something—for an academic talk it is you and the value of your research, and for an industry talk it is you and your skills. While you are selling these things to scientists, they are mostly nonspecialists in your field. You should not go overboard with self-promotion, but you also should not be self-deprecating.

Here are your principal objectives for a job talk:

- For an academic talk: to present a compelling story about your research that shows you are bringing a research program addressing important questions that will advance your field, generate external funding, and provide opportunities for student research.
- For an industry talk: to prove that you have specific technical knowledge, skills, and accomplishments related to what they need. You are selling you and your skills that will be applied to their projects.
- For both: to show that you are an articulate and responsive speaker.

- For both: to show that you are personable, friendly, relatable, and—if at all possible—funny. That is, you will make a good employee, colleague, and mentor.

The following guidelines show you how to achieve these objectives.

Organize your talk

As shown in the figure below, your talk should be structured into three segments. The first is "Tell them what you're going to tell them." This can be a single slide that outlines what you are going to cover in the talk using bullets or headings. This slide orients the audience to what is coming and tells them what they can expect to hear about. The second and longest part of your talk is "Tell them," i.e., the main content and messages, organized using the same bullets or headings you used in the first part. The third part is the summary, or "Tell them what you told them." That is when you reiterate, in one slide, the major takeaway messages.

Job talk elements, flow, and content

Start with the "why"

I have unabashedly stolen this phrase from one of my favorite leadership gurus, Simon Sinek.[42] Sinek points out that people who inspire us do not start by telling us what they want us to do or what they did; rather, they tell us why they did it or what motivated or inspired them to do what they did. He insightfully points out that the tag line for Martin Luther King's famous 1963 address at the Washington Monument was "I have a dream" not "I have a plan." The dream was what inspired him and others who shared it.

As a scientist you may be allergic to talking about what motivates your science as being a dream. But there are plenty of ways to get at the "why" without dreams.

> *Imagine a world where we could sequence the entire genome of a microorganism or even a human in ten minutes using a handheld device. Think of how we could apply that technology to help humanity. Does that interest you as much as it does me? In the next forty minutes I'm going to tell you why this interests me and how I think we can make it happen.*
>
> or
>
> *In the next twenty-four hours Alzheimer's disease will steal the minds of thousands of people throughout the world—maybe even of someone who is dear to you. It's a disease that has frustrated and challenged some of the best scientific minds of the past three decades and continues to do so today. We have more work to do. I want to be part of that work, and I want to do that work here. Let me tell you why.*

Do not think that you cannot move or inspire people unless you are working on a public health problem. How about this?

> *Almost seventy years ago, a small group of specialists got together at Dartmouth College in New Hampshire, coined the term "artificial intelligence," and laid the groundwork for decades of innovative work in learning, intelligence, and computer science. Today we are on the cusp of an even more profound advance in these fields as we complete the race to make quantum computing a reality. Why should you care about this, and, more importantly, why should you worry about it? I'll tell you.*

It would be a cold-hearted soul who could listen to these introductions and not feel inspired or at the very least want to hear what comes next. There will be people in the audience and maybe on your search committee who will not understand the bulk of your talk but who will nonetheless respond to your passion and your commitment. For them, that might be what distinguishes you from the others. Starting your talk with the "why" is a powerful way of getting their attention.

You have too many slides and most of them are horrible

It is the fundamental truth of all research talks: You have too many slides. How do I know this? I know it because since the dawn of civilization starting with the very first scientist giving their very first presentation, they have all had and to this date continue to have too many slides. This originates from the scientist's fervent desire to show during this specific research talk every piece of data they have ever generated. Those experiments you did that went nowhere? The myriad scientific dead ends you encountered? That amazing result that turned out to be a weird but totally fascinating artifact? No one cares. They do not care about the arduous journey you took to get the stellar results you will be presenting. Your talk needs to be about where you are now and where you are going in the future, so get rid of most of your slides.

This is much harder than it sounds, and you will not be able to do it yourself because you are inevitably the worst judge of what is important or interesting in your own work. You must give your talk and show your slides to friends, colleagues, and mentors and ask them to be ruthlessly honest about what you should take out and what you should keep.

It is hard to give rigid rules about how many slides you should have in a forty-minute talk because you may spend five minutes talking about some of them and others only thirty seconds. The place to start is two minutes per slide. Much shorter than that and your presentation starts to look like a disco club light show; any longer, people will start drifting. That gets your presentation to about twenty slides for a forty-five-minute talk, a number that likely puts those of you who have already put your presentation together into a panic.

Remember that your research talk is one of several opportunities for you to highlight your work. The committee will have read your résumé, your research statement, and (hopefully) your bibliography if they requested it. You would not be here giving a talk if they did not already believe that you were a credible candidate with a productive track record and a promising future. What you need to do in your talk is to *highlight facets of your work that will be of interest to and understood by a broad audience.* This is not the time to convince anyone that you are intimately familiar with a broad array of esoteric techniques and terminology that will be beyond the comprehension of most in the room.

There are those who will read the above and think "No problem. I'll just jam all the content from my sixty slides into twenty-five slides." Bad idea. This will result in a slide deck in which each slide looks like its own poster session. The result will be incomprehensible and headache inducing. Let's go over what your slides need to look like.

GENERAL GUIDELINE FOR YOUR SLIDES

1. Your first slide should contain the title of your talk, your name, and current institution.
2. Every slide should have a conclusion on it—preferably in the title. Most research slides do not follow this rule, to their detriment. The reason for doing this is that most people's attention will drift as you talk about a slide. They will daydream, check their phone, or scratch an itch. Then they will return to your slide but will have missed what you said about it so they are lost and they will drift again. If you have a declarative title on the slide, such as "Monoclonal antibody TX3 blocks viral replication in vivo" or "High frequency sound disrupts geo-location in avians," they know what point you are trying to make and may be able to pick up the thread even after drifting.
3. Use big fonts—make sure every word is visible to the viewer sitting furthest back in the room. Numerous sources suggest that sans serif fonts are easier to read and appear "more professional" than serif fonts.[43]

4. Use more images than words. It is OK to have word prompts on the slide but avoid text narratives.
5. Pay attention to font consistency—use the same font and do not use too many colors.
6. Avoid distracting background designs on slides.
7. Minimize graphics and clip art.
8. ALWAYS label axes in a way that everyone will understand and show units of measure. This is my number one pet peeve. A graph with an axis label reading "SM-rpt ($Y^{-1/}\mu$)" is an insult. Try instead "Photon energy (MeV)." Also, make sure your axis labels are readable from the back of the room because, right now, they are not.
9. Avoid jargon, acronyms, and abbreviations because most people will not know what you are talking about. It amazes me how often I tell this to a scientist, they tell me they will do it, and then they come back the next day with their slides still full of acronyms. This is called acronym blindness. They use them with their cohort and peers so routinely that they have forgotten they are using them. Even commonly used acronyms like "SSRIs" (selective serotonin reuptake inhibitors), which have become part of our daily language, should be used with caution. Even acronyms that are universal shorthand even in undergraduate biology labs (SDS-PAGE for sodium dodecyl sulfate polyacrylamide gel electrophoresis) may give some in the audience pause. This is an area where even your friends and colleagues may not be able to help you since they suffer from the same blindness that you do unless they are in a completely different field. Find someone like that to look at your slides.
10. Builds: to use or not to use? A build is when you reveal elements sequentially on a slide, most often text in a bulleted list. If you have a word slide—a bulleted summary, say—it is a good idea to use a build to present it. If you present the whole bulleted list all at once and then read it aloud, people will read the words on the slide at their own pace and not listen to what you are saying. This is yet another way to lose your audience. Builds can also be effectively used in slides with more than one graphic element to keep the viewer's attention focused.

Below is an example of a slide with a poorly drawn (completely bogus) graph, followed by a much-improved version. Here are a few things that are wrong with the slide:

- The title is incomprehensible.
- The axis labels will be meaningless to most (actually, in this case, to all, since I just made them up).
- The fonts in the (mostly unneeded) comments as well as the axes and title are too small to be read from the back of the room.
- There is too much information on it.

In the improved version we have:

- Added an easily understandable title (to a physicist, in any case) that states a conclusion. When viewers read the title, they will know most of what they need to know about the slide and will not get lost when they drift.

- Added easily understandable words (not symbols) on the axis labels, with units, both in large fonts. Even if you are using what you think will be nearly universal symbols (e.g., T for temperature), you must explain them, e.g., "T (temperature)" or just use Temperature and forget the abbreviation. T can mean a lot of other things (Time, Tension, Tesla, Thymine).
- Eliminated the red dataset and other extraneous elements that distract from the main message.

Original:

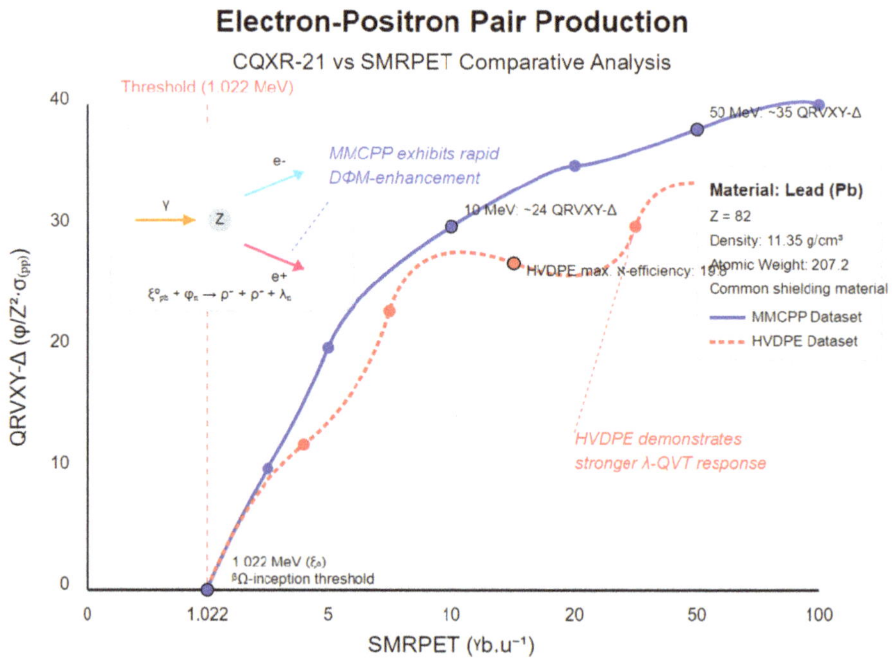

Improved:

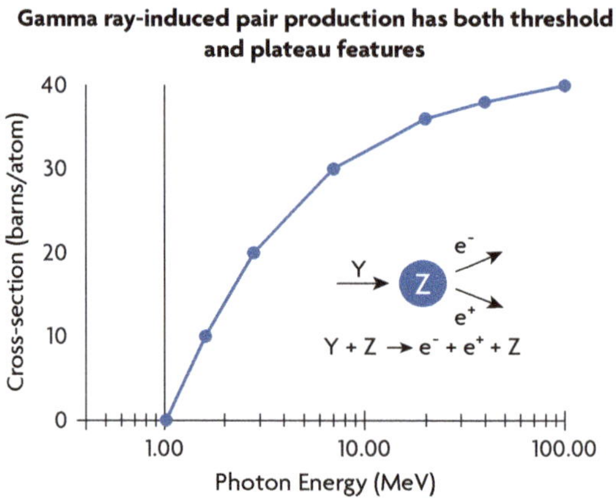

- The astute will notice that we have also fixed what was perhaps the most egregious scientific error of the original graph—the x-axis scale was neither linear nor logarithmic but rather some weird hybrid that made no sense. This has been corrected in the improved version. I took some liberty with the data points, but the graph is bogus in any case.

PREPARING FOR YOUR TALK

1. Always find out the following from your host long before you give your talk:

   A. Who will be there. This will help you to calibrate your content and level of technical detail. In an academic setting, ask if students will be there. They may be asked for their evaluation of your talk so you will need to make sure you cover highly specialized topics in ways that are accessible to them. In industry, ask if nonscientists will be there so you can adjust your content accordingly.

   B. How long you will have. If they say you have an hour, plan your talk so that you speak for no more than forty-five minutes and not one second longer. This ensures you will have time for questions, during which you will be under scrutiny. If you exceed your allotted time, everyone will be annoyed, so you need to practice your talk until you are tired of doing it. In fact, being tired of practicing is the only reliable indication that you have practiced sufficiently.

   C. Find out what kind of room the talk will be in. Try to get a look at the room well before your talk so you can be familiar with the setup.

   D. Find out from your host how this group behaves during job talks. Do they ask questions during the talk, or do they wait until the end? Knowing this will keep you from getting off balance by interruptions. No matter what they tell you, it is a good general rule to plan for interruptions during your talk. Practice being able to give concise answers and then jumping back to where you left off.

2. Well before your talk, negotiate some time with your host to test out whatever presentation equipment they have. It does not matter whether you are at NASA or Pine Gap Community College, you must make sure your laptop will talk to their projector, your USB drive will be read by their laptop, their projector actually works, the sunlight from the adjacent windows is not going to make your slides invisible, and you have a working laser pointer, unless they have a backlit screen, in which case your laser pointer is useless, so you have certainly brought your Logitech virtual pointer as well as an actual physical collapsible pointer just in case. Do you plan to write something on a whiteboard? Is there one? Are there markers? Even if there are markers, most of them will be desiccated and useless, so you have brought your own, right? And did I mention to ask whether their AV tech support is there or on call?

3. Be cautious about embedding content in your presentation that requires web connectivity, which you may lose midway into your talk. Better to have such content on your laptop. Assume that presenting anything with audio will be near impossible.

4. Do not fall into full science mode during your talk by talking too fast, not making eye contact with the audience, talking to your slides and not the audience, and being dismissive of what you think are naive questions.
5. As with your application materials, customize your talk for every organization based on what they are interested in. You may not be able to show each one a unique set of data, but you can tweak what you do show to align with their interests, needs, or focus.
6. For an academic job, showing how your work will lead to future research opportunities (both for you and your students) and will advance the field you are in is essential. For an industry job, you can highlight those ideas, but leave out your plans for doing them, since you will not be.

Job talk checklist

- Trim your slide deck
- Check out the room (lighting, layout) before the talk
- Test your laptop with their system
- Ask where their tech support person is and for their phone number
- Bring backup presentation options (USB thumb drive)
- Bring your own laser pointer and a small physical collapsible pointer as a backup
- Bring at least one dry erase marker of your own
- Find out if they ask questions during or after the talk
- Find out the composition of the audience

GIVING YOUR TALK

1. **Relax.** Begin by greeting the audience. Smile and count to three before saying anything. Transform nervous energy into enthusiasm. Pause on occasion. Allow yourself and your audience a little time to reflect and think. Do not race through your presentation and leave your audience, as well as yourself, feeling out of breath.
2. **Greet some of the audience members** as they arrive, even if you do not know them. Walk up to random people as they enter the seminar room, shake hands, introduce yourself, and ask what they do there. It is easier to speak to a group of friends, even new or momentary ones, than to strangers.
3. **Look for a smiler.** As you scan the audience during your talk, find someone who looks interested, is nodding in agreement, or is smiling. Periodically look at this person during your talk. The visual impact of an interested smiler, even in a sea of scowlers, will relax you.
4. **Scowlers are not scowling at you.** People adopt all sorts of facial expressions and body language during talks that have nothing to do with you. They may be stressed over some personal issue or have indigestion. If you see a scowler, remember, *it is not about you.*

5. **Speak with conviction.** If you do not believe what you are saying, why should they?
6. **People want you to succeed.** Audiences want you to be interesting, stimulating, and entertaining. They are rooting for you.
7. **Do not apologize** for nervousness or problems—the audience probably never noticed it.
8. **Avoid weak words** like hope, aspire, try, or seek. Use these instead: will, do, examine, study, conclude.
9. **It is about *them*, not about *you*.** Listen to questions, *respond* to their reactions, adjust, and adapt. If what you have prepared is obviously not getting across to your audience, change your strategy midstream if you can.
10. **Watch the clock.** If you are running short of time, have a plan for what to leave out. If you have extra time, have a plan for what to elaborate on or add.
11. **When asked a question, repeat it for the audience, in your own words,** unless there are only five people in the room. If you cannot repeat it in your own words, you probably do not understand it, in which case ask them to clarify it.
12. **Stop saying "That's a great question"** after *every* question. Vary it with "OK, good," "Yes, I'm glad you asked," or even "I knew someone was going to ask me that" for an especially challenging question (see the following bit on the power of self-deprecating humor).
13. **Answer questions thoughtfully.**[44] If you cannot answer it, say so and ask if you can follow up with the questioner later.
14. **Never answer more than what was asked or a question that was not asked.** Answer just enough to straighten it out, then carry on with your presentation. Continuing to talk after you have answered the question means you will be answering questions they did not ask and likely do not care about. Some are prone, when answering a question, to digress: "You think that's a problem? Let me tell you about something else that you didn't ask about that's a bigger problem." Squelch that tendency.
15. **A question that you will answer later in your talk?** Do not be afraid to say "Great point and thanks for setting up my slide that's coming in just a minute—just wait two slides."
16. **A question that requires a long answer?** Say "Good point. The answer is a bit complicated, but I do want to address it. Can I come back to that near the end?" or "I'd be happy to go through this with you after the seminar—the answer is a bit in the weeds."
17. **If you really do not know the answer to a question?** Say "That's a great point. Honestly, I don't have the answer to that at the tip of my tongue. Can we discuss this after the talk?" Do not feel like you have to make up answers on the fly—you are only human, and you cannot have thought of everything.
18. **If the questioner disagrees with you and it looks like there will be an argument, defuse the situation.** "Maybe we're looking at this differently (or 'Maybe I'm not understanding the question')—can we talk about this later? I definitely want to answer your question if I can." Similarly, if you answer a question and you can see that they are not satisfied with your answer, find out if you did not understand their

question. If it turns out you misunderstood it, it is OK to go back and answer the actual question. If they are just dissatisfied with your answer, suggest you circle back with them after the talk.

19. **Let them know when you are ending.** As you get to your last slide, say "This is my final slide ..." or "That completes my presentation. I am of course happy to take questions, and I'd like to thank you all for your attention." Or, after your last slide simply say "Thank you" and then stop talking and let them burst into wild applause. On occasion I have been on my last slide and before I can say this someone jumps in and starts asking a very complex end-of-the-seminar-type question that feels like it is going to take an hour to answer. In this case you need to regain control. "Great question. Let's just hold that thought for a second. I just want to say before taking this last question and any others there may be that I really appreciate the opportunity to speak with you all today and thanks for your kind attention" (wait for applause). Then say "Now let me answer that question ..."

During your talk, they are not only listening to what you are saying but also watching you. They are forming impressions about you that will influence what they remember about you. And whether you are aware of it or not, you are constantly sending messages about your state of mind via your body language. Indeed, it is during your talks when all eyes are on you that your quirky habits, postures, and facial expressions will be most noticeable. It is a lot to think about while you are focused on making a good impression as a scientist, but you need to remember that they are buying the whole package. These are things you can do to convey the most positive message with your body language:

Job talk body language

1. **Make eye contact.** Establish a bond with the audience.
2. **Express emotion with your face.** Smile and look surprised at appropriate places in your talk.
3. **Avoid distracting mannerisms.** Have a friend watch as you practice or record yourself and look for behaviors such as fidgeting, twitching, lip biting, key jingling, hair twirling, putting your hands in your pockets or behind your back, and making repetitive gestures. Watching a speaker make the same gesture over and over during a talk is distracting. Hardly any of us are aware of the strange things we do when giving a talk. The greatest gift someone can give you is to watch your stand-up delivery and instead of listening to what you say, only watch how you move and hold yourself, then tell you what they saw. This will be a combination of embarrassing, revealing, and baffling (as in, Why has no one ever told you this before?).
4. **Act it out.** Highlight action verbs and look for ways to act one out. Talking about neurons? Hold your hands out and jiggle your fingers like dendrites!
5. **Vary your speaking cadence and volume of speech.** Nothing gets people's attention like slowing down or speeding up your delivery to emphasize a point. Same with voice volume—moving close to the audience and saying something slowly and quietly adds emphasis.
6. **Move around.** Vary your position by moving around the room. Never remain bolted behind a podium—that is the epitome of negative body language. Walk to the other side of the room as you change topics or move toward the audience as you answer a question. Have someone observe how you do this in a rehearsal to make sure you do not have a habit of pacing back and forth.

If you cannot wow them with your science, the least you can do is entertain them with your delivery. You may get points for trying, and it will definitely enamor you to the students in the group who have to pinch themselves to stay awake during some of their professors' lectures.

Use humor (with care)

People like people who are funny. It makes them fun to be with and lightens the burden of tedious work situations. If you are not blessed with a stellar sense of humor, you will survive. But if you can make light of difficult situations, find humor in everyday events, and especially in yourself, then you may garner unconscious points in the minds of your audience.

1. Humor will enliven your presentation, can diffuse tension if something awkward happens, and helps you connect with the audience.
2. There is no sin in being mildly amusing instead of fall-down funny.
3. The safest humor has to do with yourself and is least likely to offend others. Find humor in your own life or, even better, in the moment—like when you spill your coffee or forget the name of your host.
4. Never make jokes about politics, social issues, other people's failings or foibles, sex or sexuality, or someone's appearance. Safer topics are weather, sports (except in Boston where it is no joke), air travel, traffic, and pets (except cats!).
5. If you are humor-challenged, AI can help. I asked ChatGPT to write some jokes about physics. Among the mostly unfunny options was this gem: "Did you hear the one about the physicist who discovered an elementary particle that had a half-life shorter than his attention span?"

Mistakes

Try to avoid these pitfalls when giving your talks:

1. Reading from a printed document or screen. This must never happen.
2. Failing to show passion or excitement. Especially for a job involving teaching, this can cost you points.
3. Failing to prepare and practice your talk until you are tired of giving it.
4. Failing to get frank feedback on your practice talk from anyone you can get to listen to it.

CHALK TALKS

For an academic job, chalk talks can be the most important determinant of getting the job.[45]

My own informal polling suggests that most academic interviews and probably all R1 institutions will require a chalk talk. These are an event that job applicants are least prepared for, since there is no other situation in a postdoc's experience like it. So, what are they?

A chalk talk is when you spend twenty to forty minutes outlining on a whiteboard (so no slides) your vision and plans for your research program, and typically only the search committee is there. While the talk may feel on the surface like an informal event, it is anything but, and there is a lot at stake. Also, it is another opportunity for them to see how you present and how you think on your feet.

Unlike your research seminar, during a chalk talk you should expect to be interrupted and even challenged, especially by more junior members of the search committee who may be eager to show the more senior people how clever they are. As with your seminar, practice is your friend. Practice your chalk talk with people who will ask tough questions so you can get used to answering and then picking up the thread of where you left off. My favorite quote about chalk talks is from Keith Kozminski: "Describing how to prepare a chalk talk is analogous to explaining how to ride a bike. One can describe the mechanics and steps of bike riding with considerable precision, but riding a bike is not intuitive and takes practice to keep the bike upright. Therefore, you will need the perspectives of others."[46]

Tell your host you need ten minutes in the room before your talk so you can write a few key elements on the whiteboard before people arrive. If that is not possible, saunter over to the board the minute you enter the room and start writing anyway. This is so you have a visual guide to what you will cover. A good strategy is to place these elements so that they occupy geographic domains on the board, and then write a few brief bullets as you cover each topic. Think of using the whiteboard layout as you would a poster session layout, except with less data and detail. These are the common structural elements of your talk to write down:

Focus (with two or three key words)
Plans (with a few key words)
Funding (several possible sources)
Opportunities (key words—students, faculty)

Place these headings on the white board as shown below:

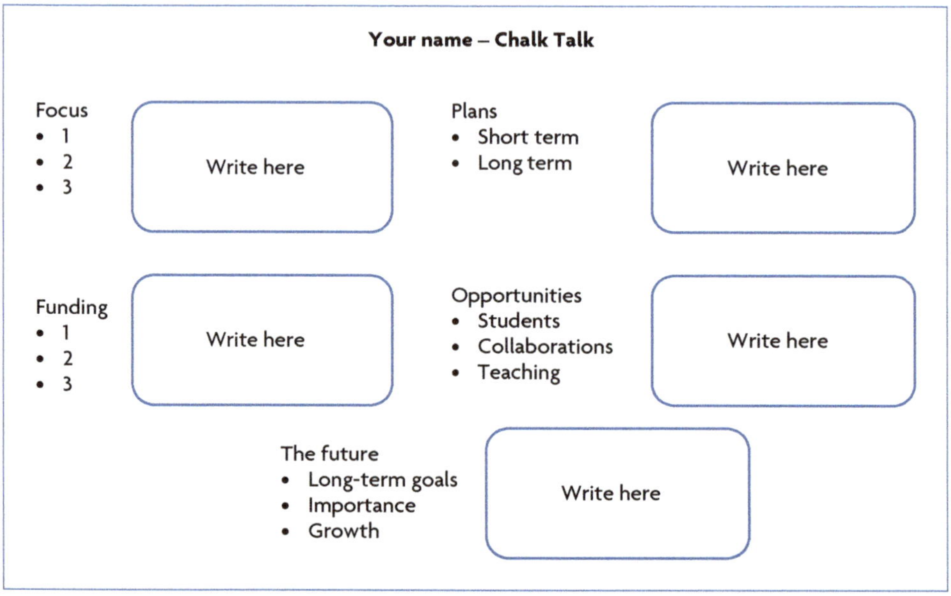

As you proceed through your presentation, write or draw key points or diagrams under or near each of these elements.

Focus

- Briefly outline your research focus, the big problem you work on, or the central question that motivates your work, and why it is important. They will likely have heard your seminar, so while you can allude to what you said there, do not repeat it word for word.

Plans

- Your near term (three to five years) research objectives and future directions with three or four aims in bullet form.
- Your longer term or aspirational objectives and where they will take you and the field. Make it clear that you understand that you will have limited personnel in your early years and that your objectives are carefully calibrated to reflect that and to ramp up as you garner additional funds and staff.
- Be prepared to address how your proposed research differs from or diverges from that of your postdoctoral work and your adviser.

Funding

- How you plan to support your work, especially as you draw down your startup funds. What agencies or foundations will you apply to? What will be the topic of your first major grant application? Demonstrate familiarity with sources of and mechanisms of funding.

Opportunities

- Mention opportunities you bring for student research—have a few ideas for research projects for graduate students who start in your lab during your first or second year.
- If appropriate, mention that there could be opportunities for collaboration with others in this department, but do not get specific about with whom.
- Summarize by highlighting the importance of your work to your field, reiterating your aspirational goals. Thank everyone for their attention and say how appreciative you are of having the opportunity to meet them.

The outline on the board ensures that you always know where you are in your talk and that you will not forget something. It is perfectly acceptable to bring a one-page printed outline of what you want to cover with you. Having this nearby and referring to it periodically will help you cover what you need to cover without having to memorize everything.

As with your seminar, pay attention to the clock. If they said you have thirty minutes and twenty minutes in you have only covered half of your outline, you need to have a plan for what to cut or breeze through. Most of the people there are very tightly scheduled and may simply get up and leave at the thirty-minute mark. This means that you need to answer questions briefly, without digression, then move on to your next item. Your committee knows you will need to control the time and will understand if you limit discussion on one topic so you can move on to the next. If they really want you to dig into something that is taking a long time they will tell you.

During your talk they are looking for evidence of a robust research program that fits thematically with or fills a gap in their department. Knowing which one of these is most important to them should be apparent to you by the time you are giving your chalk talk and will dictate how you customize your talk for this group.

Also, they are evaluating you. *Can you create and tell a compelling story about your research?* Is the organization logical and easy to follow? Can you marshal your arguments in a convincing way? How you organize and present your talks will answer these questions. *Are you passionate about your work?* The tone of your voice, animated mannerisms, and visible excitement about your work will tell them that. *Are you interested in them?* Having asked good questions during the interviews and casual conversations and demonstrating knowledge about them will reveal that. And finally, *Will you be fun to have around?* If they hire you, you are going to be around for a long time. They need to like you. Refer to our discussion of the importance of having a sense of humor.

Here are two additional job talk resources worthy of your time:

Sura SA, Smith LL, Ambrose MR, Amorim CEG, Beichman AC, et al. 2019. Ten simple rules for giving an effective academic job talk. *PLOS Computational Biology* 15(7): e1007163. https://doi.org/10.1371/journal.pcbi.1007163.

Ten simple rules for making good oral presentations. 2019. *PLOS Computational Biology* 3(4): e77. https://doi.org/10.1371/journal.pcbi.0030077.

THE TEACHING DEMO

Whether you will be asked to give a teaching demonstration will depend on the host institution's practices. Even among institutions of similar type (i.e., R1, PUI, etc.), there is little uniformity. One source reported that 34% and 37% of PUIs and Ph.D.-granting institutions, respectively, required teaching demos in addition to job talks, and 47% of Ph.D.-granting institutions said that the teaching demo was as important as the research talk.[47] Unless you ask, you will not know how strongly the place you are applying to weighs the teaching demo. Whether they tell you it is weighted equally or less than the job talk, you still need to aim at acing it.

R1 institutions typically put more weight on your research program, research statement, and chalk talk, and less on teaching, while PUI institutions will be at the other extreme. But this does not mean you should downplay your teaching demo at an R1 institution. At R1institutions there can be less emphasis on teaching in the belief (supported by some data) that teaching skills are not primary determinants of promotion.[47] While stellar teaching skills may not be the primary determinant in R1 hiring decisions, weak teaching skills could lower your chances compared to another equally qualified candidate who is an excellent teacher.

The following is a list of the most important features of effective teaching demos in order of importance.[47]

How your teaching demo will be evaluated

1. *Content is accurate*. If they ask you to teach something you are not current on, now is the time to become so. Although rare, some places may not tell you what you will be teaching until you show up.
2. *Your presentation is understandable to the students*. This is where practice and feedback will come in. You do not want to have spent hours preparing your demo only to be met by blank stares when you start presenting quantum electrodynamics to nonphysics-major undergraduates.
3. *The material is well organized*. Whether it is a twenty-minute demo or a ninety-minute marathon, your material, even if brief, should have a logical progression.
4. *Topics are introduced in a way that connect to the audience* (you ask them questions, demonstrate the relevance of the topic, etc.).
5. *You are knowledgeable about the topic*. You do not need to be the world's expert on this, but you do need to be *an* expert.
6. *You are enthusiastic*. They will not know this unless you show it in your voice and body language.
7. *You are confident in your ability to teach*. Practice getting rid of the "Ahhs" and the "Let me consult my notes."
8. *You cover the appropriate amount of material for the given time and level of student*. You cannot realistically cover a lot of material in twenty or thirty minutes, especially if you want to get them involved. Focus on one or two concepts.
9. *You speak at a comfortable pace*. You will be more likely to do this if you do not try to cover too much.
10. *Your slides are easy to read*. See the earlier section on slides.
11. *You ask if they have questions* during the presentation and wait for them to answer. A common mistake is to ask if there are questions and then wait three nanoseconds before continuing to lecture. That suggests that you really do not want questions. Learn to get comfortable asking for questions then waiting for what will feel like an uncomfortably long time, which will really be only about five seconds. You will be surprised when one or two hands go up. Even better is to tell them to interrupt you while you are talking if they have questions.
12. *When asked a question, you facilitated a discussion, rather than just telling them the answer*. This is very powerful. You will be sorely tempted, especially when asked an esoteric question to which you have the answer at the tip of your tongue, to be a show-off and give them the answer. The teaching demo is the time to show that you can bring people into a discussion in a way that leads to engagement and learning.
13. *You provide a wrap-up at the end*. Remember the admonition: "Tell them what you're going to tell them, tell them, and tell them what you told them."
14. *You incorporate elements of active learning* (e.g., group or pair discussions).

You will not be able to check off all of these criteria, but you should hit about half or two-thirds of them. As you plan your demo, keep the following dos and don'ts in mind.

Teaching demo dos and don'ts

Do:

- ***Introduce yourself when you begin your demo*** and tell the audience the name and level of the course this material is for so that everyone in the room understands the context for which the material is intended.
- ***Connect with the attendees.*** Introduce yourself to some or all (if there are just a few) of the students as they enter the room and find out a bit about them. Consider bringing a pack of index cards and a couple of markers with you. Hand out the cards and have them write their names on them to use as mini-name tents, so you can refer to them by name when they ask questions. Small thoughtful acts like this will endear you to the students. When you speak, look at the students, not the slides or board. The demo is more about their engagement than about information transfer.
- ***Make sure that what you do in your teaching demo matches what you put in your teaching statement.*** If you extolled the virtues of active learning in your statement and then used your teaching demo only to lecture, they may correctly conclude that your statement was written by AI.
- ***Show enthusiasm for what you are teaching.*** Find something in your content that is surprising or exciting and show that in your affect and words. Students engage when the teacher is engaged. Some will be asking themselves "Would I enroll in this course based on this demo?" as a way of evaluating you.
- ***Find ways to get them involved.*** Pairing up students is a common technique to involve them. Ask a question, then put them in pairs to discuss their answers. Poll the group to see how many pairs agreed and what they agreed to. Ask them to predict the outcome of a hypothetical experiment, or put multiple-choice questions on a slide and have them vote on answers.
- ***Do your homework on the topic they are asking you to teach.*** If they use a textbook, you must review this topic there to see how they expect it to be addressed. If possible, find someone you know or someone at your institution who actually teaches this material and see if they will share their lecture notes with you or at least give you their syllabus outline. There will likely be lectures or teaching modules online for this topic that you can also use.
- ***Remember that teaching science is less about transmitting information and more about helping students learn how to evaluate and use information***, i.e., how to think scientifically. You can do this by challenging them to draw conclusions from data you present or asking them how they would go about answering a scientific question you pose.
- ***Practice your teaching demo*** and ask that your audience give you honest feedback. After your first trial demo there will inevitably be much to adjust, including that you almost certainly tried to include too much. Here, less really is more. Having an in-depth discussion about one topic is better than twenty bullets about several topics.

Don't:

- Give a teaching demo that is just an extension of your research seminar. That is not what they are asking you to do.

After your demo, you may be asked questions about your teaching style or philosophy. Since you have already thought about these things while preparing your teaching statement, you will be ready. Here are some sample questions to prepare for:

- Describe the teaching philosophy you just used. Be prepared to discuss a specific tool or technique. "One thing I did was to use collaborative learning—for example, when I asked the students to discuss with the person next to them how they could test whether …"
- How would you assess student learning for the topic you just presented? Have a few ideas that do not involve a final exam. Do you use quick multiple-choice-question slides at various points during class? Do you review and ask questions about the key messages of the previous class at the beginning of the next? What kind of take-home assignments do you assign? Do you use team-learning exercises during or between lectures? Are there labs that go along with this topic?
- Why did you use (they will mention some strategy or tool you used in your demo)? This gets back to the need for consistency between your written teaching statement and what you just did. Your teaching statement and your demo need to be an integrated package, just as your research statement, your job talk, and your chalk talk need to be an integrated package.

The weirdest teaching demo is being asked to teach to a group of faculty with no students. How this is supposed to reflect your ability to engage with students is anyone's guess. Situations like these will make you wonder about the culture of the place and whether it is the place for you.

CHAPTER 8 SUMMARY

Three main performance components

- Research talks (job talks), chalk talks, and teaching demonstrations must be consistent with written research and teaching statements to create an integrated package.

Research talk (job talk) strategy

- Understand the difference from typical research seminars.
- Most of the talk must be accessible to nonspecialists in the audience.
- Essentially demonstrate future potential and communication skills.
- Must focus on selling the candidate rather than just presenting data.
- Avoid common pitfalls: too many slides, excessive jargon, and overly technical content.

Chalk talk essentials

- Practice giving informal whiteboard presentations outlining your research vision and funding plans.
- Presented to the search committee without slides.
- Often considered the most important factor in academic hiring decisions.
- Requires preparation for interruptions and challenging questions.
- Must demonstrate independent thinking and future research direction.

Teaching demonstration requirements

- Showcase pedagogical skills through actual classroom instruction.
- Key elements: accuracy, organization, enthusiasm, and active learning techniques.
- Must incorporate techniques that match stated teaching philosophy.
- Should engage the audience and demonstrate effective communication.
- Consistency between teaching statement and actual demonstration is crucial.

Preparation and delivery guidance

- **Job or research talks:** Start with "why" to inspire audience, use two-minute rule per slide (~twenty slides for forty-five minutes), and practice until "sick of giving it."
- **Chalk talks:** Write key elements on board before audience arrives, prepare one-page outline as backup, and practice controlling time and handling interruptions.
- **Teaching demos:** Bring index cards for student name tents, prepare for twenty- to thirty-minute sessions with limited content, and practice active learning techniques like pair discussions.
- **All presentations:** Never exceed allotted time, repeat questions for audience, and vary speaking cadence and use purposeful movement.

Integration and consistency

- All presentations must reinforce the candidate's overall narrative.
- Consistency across written materials and live performances demonstrates authenticity.
- Creates a compelling, unified picture of the candidate's capabilities and potential.

9 The Negotiation Playbook

Data, Tools, and Tactics for Job Offer Success

[Negotiating and going away friends]

Don't you feel guilty asking for so much startup funding knowing that there are others in this department who got only a fraction of what you are asking for?

This question was asked of a candidate discussing their startup package with the chair of an academic department in a highly regarded R1 research institution. There is so much wrong with this question that it is hard to know where to begin, yet questions like this get asked of candidates with some frequency. Your ability to advocate for yourself, your knowledge of the norms and standards of startup funds, and your confidence in the value you bring to the institution will all play a decisive role in how you handle such questions.

YOU MUST NEGOTIATE

This chapter outlines the basic tools you need to be an effective negotiator. Since the parameters and practices of negotiations differ somewhat between academia and industry, the *use* of these tools in each context is covered separately in Chapters 10 and 11, respectively.

The likelihood that the first offer you get for a position will be so astoundingly generous that you would be a fool to say anything other than yes right away is essentially zero. Even though they would rather you did not, they expect that you will negotiate and will likely think more highly of you when you do. Negotiating the elements of your offer demonstrates that you have confidence in your value, especially for mid- to high-level positions. It also shows that you can advocate for your interests in a principled and professional manner, which is an essential skill that organizations seek in candidates at the managerial or supervisory level. The above question asked by the chair is essentially a test of your ability to do this. In this chapter you will learn how not to be manipulated into thinking you are being greedy or selfish by asking for what you need to succeed.

The news that you are being offered the job will typically come by phone, or perhaps via a scheduled video call. It begins what can be a series of pleasant conversations or the most complex part of your application process—negotiating an agreement. Most hiring managers, department chairs, or deans do not want to create an adversarial or manipulative environment. After all the effort they put into deciding to make you an offer, they do not want to alienate you or give their institution or company the reputation of being difficult or manipulative. You should start by assuming that they will deal with you respectfully and honestly, and you should do the same.

While it may look like they have all the power, it is not that simple. Take an academic search as an example. They formed a search committee; they met and hashed out what kind of person they are looking for, and it was not easy—there was a lot of disagreement before they could come up with something everyone could live with. Then they went through the time-consuming process of reviewing résumés—maybe hundreds—doing screening interviews of twenty candidates, and long on-site interviews of four of them—all punctuated by long committee or team meetings. After all of that, they have decided that you are the one they want to make an offer to. Do you think they really want to move on to their second choice—if they even had a second choice, which often they do not—because the dean or hiring manager cannot reach an agreement with you? If you answered "no," go to the head of the class. So, you do have power. In the following, we will show you how to exercise that power to get the best job offer possible.

In the following you will learn how to:

- ***Get as much data as you can.*** Do your homework. Identify the parameters of what you are negotiating, including what is negotiable, what the ranges of the parameters for those items are, and what precedents or standards exist.
- ***Set targets.*** After identifying all the negotiable parameters and ranges, identify three guideposts: what your ideal package looks like, what a package you could live with looks like, and what a package that is the least you will accept looks like. This is not as complex as it may seem because while there are many elements to a job offer, most will be easy to agree on. There will likely be only three or four major elements that will require serious negotiation.
- ***Understand the basics of the negotiation process.*** A few simple tools will get you through 90% of situations.
- ***Keep yourself focused*** on the process and the outcome at all stages of the negotiation. Do not personalize the process by getting annoyed with the person you are negotiating with.
- ***Be appreciative and thankful*** throughout the process while keeping in mind that everything is hypothetical until they send you a written offer and you have accepted

it in writing. Never say "yes" or even "sounds good" until you have seen it in writing and confirm that it is exactly what you agreed to.

First Principles

- *Wait until they make you an offer. Then start negotiating.*
- *Before the offer comes, do not mention specific numbers—only ranges, and then only if pressed.*
- *Never say "yes" to anything until you see it in writing.*

Just as important as it is to negotiate, it is equally important *how* you negotiate. How you comport yourself during a negotiation, how you approach differences of opinion or viewpoints, and how you work to bridge differences show them who you are as a person. Listening, showing respect for their concerns and interests, and seeking agreements based on principles rather than on being manipulative are all signals of what kind of employee or colleague you will be.

Unlike negotiations where the two parties want dramatically different outcomes, in a job offer negotiation you share the same goal and desire the same outcome. They want to hire you, and you presumably want the job. Moreover, they want you to be successful once hired, and so do you. These shared goals, in the language of negotiation, are your shared underlying interests. They are the basis of your negotiation strategy.

The central underlying principle behind the negotiation is a simple one:

The shared goal of your negotiation is for you to have the resources and compensation that will allow you to succeed at what they are hiring you to do. They have as much at stake in your success as you do.

For academic positions they are investing startup funds in you, maybe bringing you in as their only faculty hire in the department that year; for a company position they are entrusting you with the advancement of a key company program or putting you in charge of a team of a dozen highly paid scientists. In both cases, their stake in your success is enormous. They will understand your need to create initial conditions that maximize the chance of your and their success.

If you follow the guidance in this chapter, you will be ready to negotiate the terms for any job you are offered. The keys are to gather data, learn a few negotiation tools, and then practice.

Data + The right negotiation tools + Practice = Complete serenity

GATHER DATA

When negotiating an offer, there can be considerable anxiety on your part, and for good reason. First, you probably have little or no experience negotiating in general and negotiating job offers in particular. By contrast, the person you are dealing with—the hiring manager, recruiter, or HR representative in industry, or the department chair or dean in academia—likely has a lot. Second, in addition to their experience, they have a lot of information that you do not have. They know what they can offer you in terms of salary and benefits, and they know what others like you receive for similar jobs in their organization. If you have not done your homework, and even if you have, you may only be able to guess at these things.

The only thing worse than negotiating without data is not negotiating at all. You need to do everything possible to understand not only what is negotiable but also the range of values or characteristics for each element you will be negotiating. Here we will cover how you can gather information on some of the most commonly negotiated elements. The next section covers all of the items that are typically negotiable in job offers in academia and industry.

There is considerable benefit in asking people who interview you about specific negotiable items related to your job, whether it be startup funds or teaching load. These people are your best source of information since they have specific and presumably accurate data. The key is how and under what circumstances to ask.

If you are a candidate for a junior faculty position being interviewed by a faculty member, they may be willing to share insider knowledge with you. They may tell you about norms for startup funds and teaching expectations as a way of showing their friendship or collegiality or helping you through a situation that they themselves have gone through recently or to support your candidacy. The more firsthand information you gather, the more reliable data you will have when the time comes to negotiate a final offer or make a decision. Near the end of an interview day or process, asking questions of the hiring manager, dean, or chair about the parameters of a potential offer is perfectly acceptable.

Remember that you are just gathering information. If numbers are mentioned by someone in authority during an informal discussion, that is not the time to say, "That sounds great." All you are doing is looking for guidelines to use when the real negotiation begins.

Salary is a special case. For an academic job, you should be able to gather data on salary ranges for the job you are applying for from publicly available sources, or the range will have been specified in the position posting. While everyone knows that salary is important, there is no reason to signal that it is the first thing on your mind. For early career faculty or industry positions, salary may be one of the least negotiable items, and, if negotiable, only within a very narrow range. Asking too early in the process may not yield much useful information beyond a parroting of the posted range.

More and more organizations, including the federal government, post salary ranges when advertising open positions. In many states, such transparency is required by law. The downside is that ranges are often exceedingly broad with multiple, indecipherable levels so that figuring out where you might land in the range is a guessing game. Publicly available sources worth studying include the following:

- The Chronicle of Higher Education[48]
- Inside Higher Ed[49]

- American Association of University Professors[50]
- H1B Salary Database[51] (excellent salary database searchable by position and organization)
- Indeed[52]
- Randstad salary guide[53]
- Glassdoor[54]
- Bureau of Labor Statistics[55]
- Yale University's Office of Career and Professional Development[56]
- Future PI Slack[57] (lists job search resources and has an excellent spreadsheet current through 2023 with useful firsthand data on job offer experiences)[58]

Some states, California and Illinois among them, make the faculty salaries at their state university systems public via a searchable database. The University of California system is especially useful as it is comprised of some of the most highly rated universities in the country.[59] While you would be remiss not to spend time looking through these sources, be aware that in these dynamic economic times for science and education, things may not be now what they were when the data were gathered—typically a year or more in the past.

In the private sector, getting information about salary ranges can be more challenging, since private sector organizations do not make their salary guidelines public. You may be able to find data on specific industries or companies on sites like Glassdoor. If you are applying for an entry-level position and you are meeting individually with a potential peer or teammate, you could ask very generally about salary, as long as you are not asking them about *their* salary. Asking this question to a group of potential peers or teammates is not recommended, as they will be unlikely to share such information among themselves.

A NEGOTIATION TOOLKIT

Negotiating your job offer will not be a single event or conversation but rather a process that will take place over days or even weeks depending on the level of your position. Learning how to be an effective negotiator can make the difference between obtaining a satisfactory offer that you are happy with and a skimpy one that leaves you feeling that you have been taken advantage of. Moreover, whether in academia or in industry, learning how to negotiate skillfully will be a major asset throughout your career. In the following section we outline some of the key elements of negotiation specifically as they apply to negotiating a job offer, but the principles apply to other circumstances as well.[60]

Tactical tools

1. Prepare for it: Set targets
2. Underlying interests: It's not what you want, it's why you want it

3. Standards and criteria: What do you mean by "more"?
4. Open-ended questions open doors
5. Don't get personal: Focus on the problem, not the person

Behavioral tools

6. Behave yourself: Manage your behavior
7. It's your body talking: The power of body language
8. Be an influencer

Prepare for it: Set targets

If you have been successful at gathering data, you will be going into the negotiation knowing the likely salary range, the possible types of benefits, and the range of startup funds that could be available to you. This information will allow you to calibrate your expectations. If you have not been able to gather such data, you will have to make educated guesses based on data from other organizations or anecdotal information. In either case, before the negotiation, carefully think through your needs and expectations and write down the following three outcome scenarios:

- ***An ideal outcome.*** What would an agreement that satisfied all of your requirements look like? If you could get everything you wanted in terms of salary, startup funds, and benefits, what would that look like for each?
- ***An acceptable outcome.*** If you could not get everything that you listed in your ideal outcome but could get an offer that made you happy, what would that offer look like?
- ***The least outcome that you could accept.*** This is your bottom line. What is the least you would accept in an offer, anything less than which would cause you to walk away?

In each of these scenarios there will be multiple parameters to evaluate. For an academic position these include salary, startup funds, teaching responsibilities, and lab space. For a job in industry, they include salary, stock options, position title and responsibility, and potential for a bonus. To manage the complexity, list each of the major negotiable elements and write down your ideal, acceptable, and bottom line for each. Creating the three hierarchies of potential outcomes helps you know where you stand as you progress through your negotiations. Are you getting closer or further away from your ideal or satisfactory outcome?

Underlying interests: It's not what you want, it's why you want it

Always negotiate on the basis of *interests* rather than on the basis of *positions*. A position is something you want: "I need $2,225,000 in startup funds." Your interests are all the reasons why you need this amount of money. Framing your request for a specific amount of startup funds on the basis of interests looks like the following:

I've put together a detailed budget of my needs during my first three years, including lab staff salaries, equipment, and supplies until such time as I can generate a significant amount of funding on my own. That leads me to conclude that I need a minimum of $2,225,000 in startup funds in order for me to be successful in starting a program in immune regulation, start training students, and put myself in a position to get my first independent funding award.

That may seem like a lot of words, but you are asking for a lot of money. You are backing up your request with a defendable rationale. That rationale is your underlying interest—that they and you want you to be successful. Everything you ask for, whether it be startup funds, space, or salary, needs to be grounded in this fundamental underlying mutual interest. Going about it any other way makes it sound like you are just asking for more.

> *For a faculty position, you need to be able to justify everything you negotiate for by the underlying interest that you both share: the success of your research program.*

Being able to justify your requests in this way makes you immune to the kind of aspersions cast by the department chair at the beginning of this chapter, i.e., that you are just being greedy.

***Always* justify from interests**

Position	Interest
Salary	Need to be compensated fairly in line with expertise and experience
Equipment list	Need to be productive quickly
Limited first-year teaching and committees	Need time to write grants
Support for postdocs/students	Need to get started with research to increase likelihood of funding
Support for meeting travel and registration	Must keep current with my field and represent this institution

There are other interests that you have beyond the success of your research program. These include being treated fairly. You and they need to agree on a fair salary, one that is commensurate with your background and experience and consistent with what others with the same experience and in a similar role receive. That could be the basis for asking for higher compensation than they are offering you.

Of course, interests are not always shared. You may believe that you are worth more money than they feel they can pay because they claim to have an interest in ensuring that people in the same position are paid comparable amounts, and what you are asking for is outside this range. *However, organizations routinely compensate people they really want with a higher salary and other perks than they publicly admit to.* If they really want you, do not let their appeals to fairness, equity, and internal standards dissuade you from negotiating what you think you are worth. If you really do wish to push the matter, you will need to risk their concluding that you are too expensive, in which case they may withdraw their offer. Doing research, especially finding out what kinds of offers people like yourself have received at this or similar institutions, is the key to judging how much more you can ask for without pushing them to this conclusion.

If they insist that the reason they cannot offer more has to do with their interest in preserving what they see as pay equity, you need to pivot to a different tactic, using standards and criteria.

Standards and criteria: What do you mean by "more"?

If you have done research or they have revealed the salary range for this position, then you may already know whether your number is outside the range or is simply at the highest end. If it is the latter, then your job is an easy one—you need to find standards or criteria that put you there. Those could include the following:

- You are considered a rising star in your field, have an impressive publication list, and have four other job prospects at other institutions.
- You already have external funding.
- You have more years of work experience than the average candidate or that was specified as a minimum requirement.
- You have a higher skill level in certain critical areas than the bare minimum they specified.
- You hold an advanced degree in a relevant technical discipline.
- You have highly relevant industry experience in a peer organization.
- You are skilled in a technical area that, while not specified in the job description, will add significantly to your value (i.e., you are applying for a job in the assay development group, but you also have deep experience in AI and assay automation).

Any of these can form the basis for a higher compensation that would not run afoul of their "equity" argument because it would be based on your having *valued qualifications* beyond their expectations. If it looks like this tactic is not going to help you and you are willing to start at the compensation level they are proposing, you could negotiate an early promotion review or ask whether the position could be reclassified. Here are other kinds of standards and criteria:

- Anecdotal information you may have acquired during your interview about the terms offered to others with positions similar to yours
- National data on salary ranges for this type of position (especially useful if their offer is considerably outside of the range)

What if they try to use their own version of standards by asking about your current salary? In many states, asking about current salary is illegal. But illegality will not necessarily stop them from asking, and you will need to decide how to respond, since saying

"You can't ask that, it's illegal" is not going to make the remainder of the interview very pleasant. If your current salary is in the target range of what you want for this job, there is no reason not to share it. If it is much lower than what you expect for this job, you can say "I'm not sure my current salary is relevant as the job I have now is very different from this one with far less responsibility and technical expertise required."

Open-ended questions open doors

At some point during any negotiation the other side may say "We don't usually do that," "That's probably not going to be possible," or "We can't do that." These statements are of increasing strength, but the first two suggest that this item could be negotiable although they would rather that it was not.

> *A "no" is not the end of the discussion;*
> *it's just the beginning of a different discussion.*

"We can't do that" can be their way of signaling that they really do not have any flexibility. Such could be the case if you ask for a salary that is significantly above the organization's firmly mandated guidelines. But it could also be a tactic.

Let's say you asked whether you could have a larger lab space, which you hear will be available in a building that is just about to open. Their saying "That's probably not going to be possible" could mean that it *is* in fact possible under some set of conditions. Your job is to determine those conditions. You might respond, "Well, this is pretty important to me. What would we need to do to get space in that building? (*Note that we have added them to your problem-solving team by the use of "we," making reference to your shared interests.*) My concern is that the space you showed me will not be adequate to get my program running at a level that allows me to generate my own funding in the next three years. I don't want to be coming back to you in three years asking for additional bridge funds to keep me going. That's my main concern here." Now you are negotiating again.

What if they say "We can't use that space, it's reserved for our senior faculty." This may sound like a hard no, but you should not assume that. It could mean that getting that space would be a pain in the neck for them and they would rather not even try. In response you can say almost exactly what we suggested above, but just start it a little softer, reflecting that you heard the hard no—"OK, I hear you, just out of curiosity though, what would need to happen for me to get space in that building? My concern is that ... etc."

Ask open-ended questions to keep the dialogue moving and generate new options

Use questions that start with:

- "What if we . . .?"
- "Have you considered the possibility . . .?"
- "What would we need to do to make that happen?"
- "What kind of data would cause you to look at this differently?"
- "What would be wrong with . . .?"
- "How could we get the committee to support this?"

Technically, they gave you a hard no, which you countered with an open-ended hypothetical question, backed up by justification that references your shared interests. And you did it nicely. They could still say "Not going to happen," but they may also be swayed by your justification. If you signal that this item is a real priority for you and potentially necessary (although unspoken here) for your decision to accept the offer, you could be back at the negotiating table.

Finally, if the new building really is a no-go zone, you might counter with "OK, I understand. Given the importance of this though, what other space alternatives are there? How could we find a space that would fit the program you want me to build?" Another open-ended question, involving them in finding a solution. Again, note the word usage. Not "Can we find a space …" but "How can we find a space …?"

The power of most open-ended questions is that they cannot be answered with a yes or a no. Asking "Is there any chance I could get space in the new building?" is too easily answered with a "no." But "What would need to happen for us to get space in the new building?" requires the other side to respond substantively, thus keeping the conversation going. The open-ended question takes advantage of a very powerful psychological ploy—asking for advice. By framing the question as you asking them for their help in coming to a resolution, you do not come across as forcing *your* solution on them.

Startup funds discussion

You: "The first item I'd like to go over is my startup funds" (not "There's no way I could live with these numbers").	Nonconfrontational approach.
Them: "That's basically our typical package; there's not much room for negotiation. You're not suggesting you deserve more than others, are you?"	
You: "I know you need to be fair. All I want to do is make sure I have the best shot at ramping up my research and independent funding as quickly as possible. I know we both want that. These numbers might make that hard to do. Can we go through the list I sent you and I'll explain which are the critical items?"	Be empathic. Ignore any stonewalling and dismissive comments. Don't get defensive.
Them: "I can't do anything about that. The dean's pretty rigid about these things."	
You: "OK, is there any way we can set up a meeting with the dean? All I'm trying to do is to make sure I'm as productive as I can possibly be starting from day one."	Don't accept this "appeal to higher authority." Get the higher authority involved.

Don't get personal: Focus on the problem, not the person

If you are having a hard time coming to an agreement, your frustration can spill over onto the other person. If they seem intransigent on a matter that is important to you, if they minimize your concerns or renege on something you thought you had agreed on, you can find yourself thinking unpleasant thoughts about them. There is no harm in such thoughts, as long as they do not manifest themselves in behavior. Unfortunately, unpleasant thoughts about someone have a nasty habit of leaking through into behavior more often than we would like. Making a sarcastic comment, responding to something they said with even mild annoyance, or looking annoyed during a discussion are all going to do more harm than good to your negotiation prospects.

Don't make the other person into the problem.

If you think your joint inability to reach an agreement is because the other person is being stubborn, pigheaded, closed-minded, petty, or penny pinching, then you have turned them into the problem. And now you are stuck because if *they are* the problem and you cannot fix *them* (which you cannot), then you cannot solve the problem. There is nothing to be gained by grousing about "Why can't they be more flexible?" or "Why are they so insistent on this teaching load?" If you find yourself slipping into this train of thought, take notice and pinch yourself. It will lead nowhere good. Get back to the substantive matters at hand.

Behave yourself: Manage your behavior

Beyond knowing the best tactical approach to take, being in control of how you react in the moment can also be key to achieving a satisfactory outcome. Feelings of guilt, anger, or frustration are normal reactions, but *acting* guilty, angry, or frustrated are counterproductive. There can be circumstances in a negotiation when something is said that makes you feel angry, frustrated, or disrespected. While most people have enough self-control that they would never act on these feelings during a job interview, their feelings may nonetheless be evident by their tone of voice or facial expression.

The first and hardest part of managing strong feelings is to *recognize* them as you have them in real time. Failure to do this can result in your acting on them before you harness them. If you *can* do it (it takes a lot of practice), you can work to make sure they do not manifest themselves in your affect or in your words. Have an internal dialogue instead: "OK, that made me angry. This guy's a jerk and there's nothing I can do about that. I need to maintain my composure and get back to the negotiation."

Another tool for managing yourself during a difficult conversation is "going to the balcony." This tactic described by William Ury involves picturing yourself above the meeting table or room, looking down from above.[60] Mentally removing yourself from the room gives you a different perspective on what is going on. From this mental vantage point, assess what is going on between the two (or more) of you. Perhaps you felt upset about something they said and unknowingly reacted in a way that did not result in the outcome you hoped for. Going to the balcony may allow you to get back on track—"Let me just restate that ..." or "Let's roll back the tape a bit and go back to where we were a minute ago ..."

It's your body talking: The power of body language

The aspect of your own behavior that you probably pay the least attention to, in part because it is literally invisible to you, is your body language.

Because the implied meaning of body language is so transparent, you need to pay attention to yours. It can be counterproductive if your face is expressing skepticism (raised eyebrows) or disagreement (frowning), and that is not what you want to convey. When you are frustrated with someone, you still need to get to an agreement, so showing frustration on your face or in your body (tense posture or crossed arms) without being aware of it may be sending messages you do not want sent.

Positive Body Language

Relaxed posture

Eye contact

Moving shoulders toward them

Relaxed hand gestures

Facing square toward them

Positive nod of head

Interested facial expression

Smile

Body language is just as important when meeting someone virtually as it is face-to-face as we discussed in Chapter 6 on screening interviews. That the impact and meaning of body language is every bit as apparent on video as it is in person is illustrated by the following anecdote.

THE POWER OF BODY LANGUAGE

For years I have been running workshops on negotiation skills for scientists. One of the exercises I use in my workshop illustrates the power of body language. I ask people to break up into pairs and stand up and face each other. I tell them that one person should tell the other what they did for fun or enjoyment last weekend and that while they are doing that the other person should display ***positive*** *body language. As soon as they finish, they need to tell that same person what they are going to do for fun and enjoyment next weekend during which the other person should display* ***negative*** *body language. Then I have them switch roles so that they each get a chance to be a talker and a body language displayer. At the end of the exercise, I ask people what their experience was when they were on the receiving end of the negative body language. Invariably the answers are "I felt like I couldn't keep talking," "It just felt horrible," "It felt like all the air was sucked out of the room ..." The types of negative body language being displayed were not what you would consider over the top or egregious. Some people were just looking at their watch, others started scrolling through their emails on their phone, and others simply let their eyes wander around the room—typical kinds of body language that anyone might display in almost any face-to-face interaction.*

When the pandemic struck in 2020 and in-person workshops vanished, I started getting calls asking me to run workshops on Zoom, which I had never done before. I figured it would be easy to adapt the workshops to the Zoom format, and it was. But I was concerned about whether the body language role play would translate from an in-person format to a video format where all people could see of each other might be their faces. When the time came for my first Zoom workshop, I paired people off and dropped them into breakout rooms to perform the body language exercise. In the end my concern was totally misplaced. When I ran this exercise on Zoom, it was immediately apparent that not only were people adept at displaying negative body language via Zoom but that the impact on the recipient was every bit as powerful and immediate as it was in person. Even subtle negative cues like the speaker not looking at their camera or not so subtle facial expressions conveyed messages that were just as powerful as they were in person.

What is most surprising about this anecdote is not that body language comes across on video but that people on the receiving end, whether in person or on video, had such strong reactions despite knowing that the other person *was acting*. Facial expressions and physical actions have an immediate and powerful impact on the other person. You need to routinely monitor yourself to ensure that you are not inadvertently sending negative messages.

Be an influencer

Just as you can ensure that your body language is sending the right message to the other person, you can do the same with the words you use. Your goal is to demonstrate that you are hearing what they are saying, even if they have said something that you disagree with. Ways to do this include the following:

1. **"So, let me get this straight …"** Summarize and paraphrase what they say. There is a lot at stake during job negotiations, and feeling anxious is a common response. Such feelings can cloud your thinking and distract you from what is being said. Similarly, if your mind is flooded with counterarguments to what they are saying, then you are no longer listening; you are planning what you are going to say next. The consequence will be that when you do respond, you are likely to hear back "That's not actually what I said …"

   The solution is that after they have spoken, or made a proposal, repeat it back to them. "Let me make sure I have this right; you are suggesting that …" Even if you are certain you heard them correctly, repeat it back. More often than you would expect, what you will hear back will be "No, that's not what I said …" The benefit of doing this is twofold. First, you now have accurate information instead of your inaccurate recollection. Second, by repeating their words back to them, you show that you are paying attention and care about their viewpoint.
2. **Agree whenever you can and get them to say yes.** You may find that you need to negotiate a list of items, but you do not want to come across as a nitpicker. The solution is to look for and make explicit the things you *do* agree with and call those out. "I'm glad we're on the same page with most of these items, especially the start date. I think you have been really flexible, and I appreciate that." Asking them questions to which they need to answer "yes" can put them in an agreeable frame of mind and emphasize areas of agreement.[60] Even a simple "yes" to "Is it OK if we put this item aside and move on to the next item?" or "Can we get to work on this short list? Is now a good time?" or "I think we've resolved that issue, would you agree?" works.

3. **Are you talking to me?** Using the other person's name is a method of establishing a connection and creating a momentary bond. Saying "Fred, I think we just have a few items left to discuss," or "Dean Ortega, I really appreciate you taking the time to meet with me," personalizes your comment or question. But do not overuse this. Saying someone's name more than once or twice in any conversation can sound creepy.

4. **Lose the "buts."** This and the next tool are the easiest and maybe the most powerful conversational gambits in the "Getting past no" playbook.[60] They say "I think this is a very fair salary offer we're making you and it's really as high as we can go." You could then say "Yes, *but* it looks like you are not really giving me credit for the fact that I have five years of experience doing exactly this job at your biggest competitor."

   The "yes" implies that you heard them; the "but" signals that you disagree. It also implies that they are at fault for ignoring something. While this may be true, pointing it out in this way is not your best next step. Try this instead: "Yes, *and* now let's discuss how we can take into account that I have much more experience than the typical candidate for this position."

   This is a version of our advice on dealing with a hard "no" earlier in this section. They are saying "no" to a higher salary, but you just keep on negotiating, sidestepping the disagreement. With the "Yes, and ..." tool you are pointing out that two things can be true simultaneously—*yes*, they made what they think is a fair offer, *and* at the same time it did not account for a key consideration. The "and" does a lot of heavy lifting.

"But . . ." vs. "and . . ."

Them: "We can only offer you half of the startup funds you requested." **You:** "OK, but you know that will mean I won't be able to start my most important project." vs. **You:** "OK, so let's discuss how that will impact your expectations for my research progress and funding goals."
Them: "So, this is the only research space left in the department. I know it's pretty small." **You:** "But you completely ignored the equipment that needs to go in there." vs. **You:** "I see, and can we talk about where all my equipment can go? Here's the list we agreed to."

5. **Use "I" statements instead of "you" statements.** Referring to the example we used above, a "you" statement is "You didn't give me any credit for the relevant work experience I have." "You" statements will almost always come across as being accusatory, which is not a good tactic in a negotiation. Accusations will trigger an automatic denial: "No, that's not true. I did take that into consideration," and then it is you said, they said.

   Instead, the way to express this is "I'm not sure this offer adequately recognizes my five years of on-the-job experience," or "I'd like to see something in this offer that recognizes my prior experience. How can we get there? Bumping up the salary is for me the easiest approach." You are making a statement about you rather than about them. Taking the "you" out removes the implied accusation as does substituting "this offer" for "your offer."

EVALUATE THE PACKAGE, NOT JUST THE PIECES

In any complex negotiation it seems natural to make a list of issues and tackle them one at a time. In an academic negotiation there could be as many as a dozen items being negotiated. While it is true that each of these needs to be addressed, if you negotiate each item as though it is a deal-breaker you can lose sight of your priorities. Also, you can easily end up annoying your potential employer if you come across as a nitpicker. This is where having defined your priorities for each of the important items being negotiated according to the ideal, acceptable, and bottom-line criteria comes into play. In some cases, you may find that an ideal outcome on one item balances a bottom-line outcome on another. For example, unexpectedly generous startup funding may convince you to accept somewhat less space than you hoped for. On the other hand, there may be no compensatory offers that would induce you to take a salary that is considerably below your bottom line.

Negotiate the whole package at the same time but signal which items are the most important.

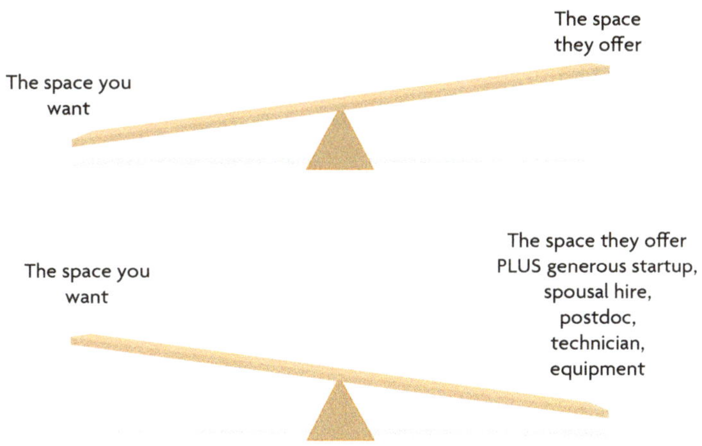

FACE-TO-FACE OR SOMETHING ELSE?

Negotiations can take place over days or weeks, requiring multiple back and forth communications via email, video, phone, or in person. Each of these modalities has its own benefits and pitfalls. Studies have found that while the success and quality of outcomes is not significantly different between email and in person, negotiators report higher satisfaction with in-person or video negotiations versus email and suggest that the "richer" the communication medium the greater the sense of collaboration.[61,62] Seeing each other enables negotiators to better read the other person's response and helps the negotiators to connect on a personal level.[62]

What seems to be important is not that *every* negotiation communication is information-rich, just that *some* are, enough to create a connection. This condition is almost always met in a job offer negation as it would be unimaginable that an offer would be extended without having had in-person or virtual meetings or, more likely, both.

But there are times when in-person or video communication may not be optimal. If the discussion has the potential to be even remotely awkward, being on the phone and not being seen can take some of the psychological pressure off.

Another advantage of the phone is that you can have notes in front of you. Being able to refer to a list of talking points, budget spreadsheets, planning timelines, lists of startup equipment, and any number of backup data sheets is something that is hard or awkward to do with any other form of communication. Watching someone shuffling through notes on a video call is immensely distracting.

Another consideration is that in-person and video conversations take place in real time. Negotiating in real time adds pressure, which can be a disadvantage if the issues are complex and you need time to think. Consider a situation where you are sitting in the vice president's office, and she says, "So, based on what we've discussed so far, will you accept our offer?" but you still have a number of concerns that have not been addressed. You could mention those concerns at that point, or you could hold off and ask for time to think about it. The downside to mentioning your concerns right away is that she may offer solutions to them right away. If she addresses the very things that were of concern to you, she is going to expect an answer. Then, you are in a situation where you are going to feel pressured to respond, and making a decision when you feel pressured is a bad idea.

This is a situation that calls for saying "I'm really excited about this job and feel very positively about joining your company. It's a lot to think about, and what I'd like to do is review what I've learned today and follow up by email in a day or so. Is that OK?" Now you can take your time thinking about what if anything needs to be negotiated further. You then send a very appreciative email outlining the remaining items to be discussed. It may be that these matters get resolved by email or by video call.

Moving from real time to email slows things down and gives you the time you need to consider your options and your response.

AN OFFER YOU CAN'T REFUSE

What if someone says "If I were to make you this offer right now, would you take it? We can finish this up today." If you have come this far in the book, you already know your answer, which is a nice "no." What is a nice no? It sounds like this: "Wow, that's amazing. This is really an attractive offer, and I very much appreciate the flexibility you showed on the salary. At the same time (*Remember above? No "buts" here!*), as I'm sure you appreciate, this is a really big decision. I definitely need some time to think this over and I appreciate your outlining the general terms of an agreement, which look very attractive. What would be great is if you could send me a letter with all these things written down so I can go over everything carefully, and then I will let you know ASAP. Does that work for you?"

Negotiation format	Features	Use if . . .	Don't use if . . .
In person	• Hardly ever the best format. It feels time constrained, they are looking at you, and you may not be prepared for an issue that arises. • May feel pressured to reach an agreement "right now, while we're together."	• You are a confident negotiator and/or the negotiation looks like it is going to be entirely in your favor and nonconfrontational.	You are: • Afraid they will put you on the spot • Nervous • Unprepared • Not confident in your language skills
Zoom	• Not much better than in person for all the same reasons. The only possible advantage over in person is that you can make reference to notes to yourself that are out of their sight. • Another advantage over in person is that the pressure to reach an agreement "right now" is less and meetings can be very short, e.g., thirty minutes.	• You are a confident negotiator and/or the negotiation looks like it is going to be entirely in your favor and nonconfrontational. • You want the benefit of watching their facial expression at key points during the discussion.	You are: • Afraid they will put you on the spot • Nervous • Unprepared • Not confident in your language skills
Phone	• Advantages include having notes in front of you and they cannot see you. • The disadvantage is the loss of facial and body language cues.	• You are somewhat nervous about negotiating but feel more comfortable when you have your notes in front of you and they cannot see you.	• You feel that the inability to see them will deprive you of important information about their reactions.
Email	• An advantage is that communications happen at a pace that you can control. You have time to carefully think through responses in a way that no other modality allows.	• Any form of real-time negotiation makes you nervous. • You lack confidence in your ability to respond accurately to proposals made in the moment. • There are complex issues being negotiated and you do not want to be put on the spot to agree to anything without the benefit of considerable thought and consultation with others.	• You wish to establish or create a personal connection with the other person. • You have had excellent discussions in person already and are concerned that switching to email might be seen as distancing yourself from them.

Chances are, if someone is making an off-the-cuff offer and asking for your answer right away, there are a dozen things that have not been tied down yet.

NEGOTIATION GUIDANCE FOR WOMEN

Historical data show that women in the workforce in general and in academia in particular continue to receive on average lower compensation for identical work than men.[63,64] A recent study at a US R1 university showed that when negotiating startup packages in the same institution, women fared significantly less well than men in multiple dimensions including amount of and number of years of startup funds, course release time, and lab space.[65] Kate Zernike's 2023 classic, *The Exceptions: Nancy Hopkins, MIT, and the Fight for Women in Science*, shows that this inequity impacts even the most accomplished female academics at our most lauded institutions.[66] There are likely multiple reasons for these differences including structural inequities. For some women, a learned or acculturated tendency to avoid negotiating in the face of social pressure in certain situations (for example, negotiating with a man) may be contributing factors.[67,68]

If you think your approach to negotiating needs improvement, I urge you to read a book that I recommend to my women coaching clients: *Hardball for Women: Winning at the Game of Business* by Pat Heim, Tammy Hughes, and Susan Golant.[69] The book is filled with useful insights including examples of how men and women approach negotiation differently in American culture. For example, recognizing that while women often value harmony and collaboration, negotiation can require assertiveness and strategic self-interest. Learning the "male rules" of interaction—not to mimic them blindly but to use them deliberately when appropriate—may help you achieve a more beneficial outcome.

While salary surveys show that women get offered less on average than men, proving this in any particular instance is difficult. If you think you have received an inadequate offer, information is your best ally. Research what others have received in similar circumstances and use this information (the "criteria and standards" tool) to support a counterproposal that better reflects what you are worth. Beyond that, the most powerful persuader is having a bona fide competing, and better, offer from another organization. Establishing your market value can work wonders during a negotiation.

Women continue to be subjected to workplace behaviors fueled by cultural norms and stereotyping. Such behaviors can be frustrating and infuriating. For example, if you were a woman subjected to the comment quoted at the beginning of this chapter, delivered by a man, you might well feel angry. As we advised earlier, what to do with that anger is a strategic decision. In Western cultures, women are often penalized and labeled as "too emotional" if they display anger or "too masculine" if they show aggression. Recognizing rather than acting on anger and deciding whether and how to respond are the keys. In the case cited, you might have said in a neutral tone, "What's at issue here is my ability to put together the program we both want. I can't fix what was done for others, but we can both make sure that I have the resources I need in order to succeed here." The additional challenges women continue to face when negotiating a job offer are frustrating and unfair, but you can use strategic approaches to mitigate these challenges and get the best offer possible.

CHAPTER 9 SUMMARY

You must negotiate every job offer

- Employers expect it, and it demonstrates confidence in your value. The likelihood of getting a perfect first offer is essentially zero, even for entry-level positions.

Do extensive research before negotiating

- Gather data on salary ranges, benefits, startup funds, and what is typically negotiable in your field using public databases, websites, and insider information from interviews.

Focus on shared interests, not positions

- Frame requests around mutual success rather than just what you want. Both you and the employer want you to succeed, so justify everything based on what you need to be effective in the role.

Use strategic communication tools

- Ask open-ended questions that cannot be answered with "no," use "yes, and ..." instead of "yes, but ...," and avoid "you" statements that sound accusatory. Replace them with "I" statements about your needs.

Manage your emotions and body language

- Do not let frustration, anger, or guilt show during negotiations. Maintain positive body language and composure even when facing manipulative tactics or pressure.

Never agree until you see it in writing

- Do not say "yes" to verbal offers or when pressured for immediate decisions. Always ask for time to review written terms, and choose your communication method (email, phone, video) strategically based on your comfort level.

10 The Art of the Academic Deal

Negotiating Your Worth Without Burning Bridges

TYPICAL SEQUENCE FOR AN ACADEMIC JOB OFFER

The following figure illustrates one of many scenarios for the offer and negotiation sequence.[70] There are as many variants of this as there are academic institutions, but the pattern will be similar.

When you get the first sign that they are interested (Step 1), you need to be ready. While you may have been asked about what kind of startup package you are looking for during some of the interviews you had, you hopefully answered in only the vaguest manner, such as "I'm still working on that …" or "I'm putting it together now; it shouldn't be too much longer. I want to make sure I include everything."

In preparation for a conversation with the dean or chair, they will likely ask for your proposed startup budget so they can factor that into an offer. You will send that to them (we will go over exactly what goes into this shortly) and you will hear back with an outline of an offer (Step 2). Be extremely thankful and appreciative but noncommittal. "First of all, thank you so much. I am really honored that you've selected me for this position. In very general terms, this looks like a terrific offer. Let me go over it. There are probably a few things that need adjusting, but I want to think them through carefully and make sure I don't miss anything before I get back to you. If you put this in writing and email it to me (Step 3), I could get back to you next Tuesday. Does that work for you?" Ask them how they would like to go over your comments—do they want you to email them your list? Do they want to discuss it on the phone?

Academic job offer negotiation sequence

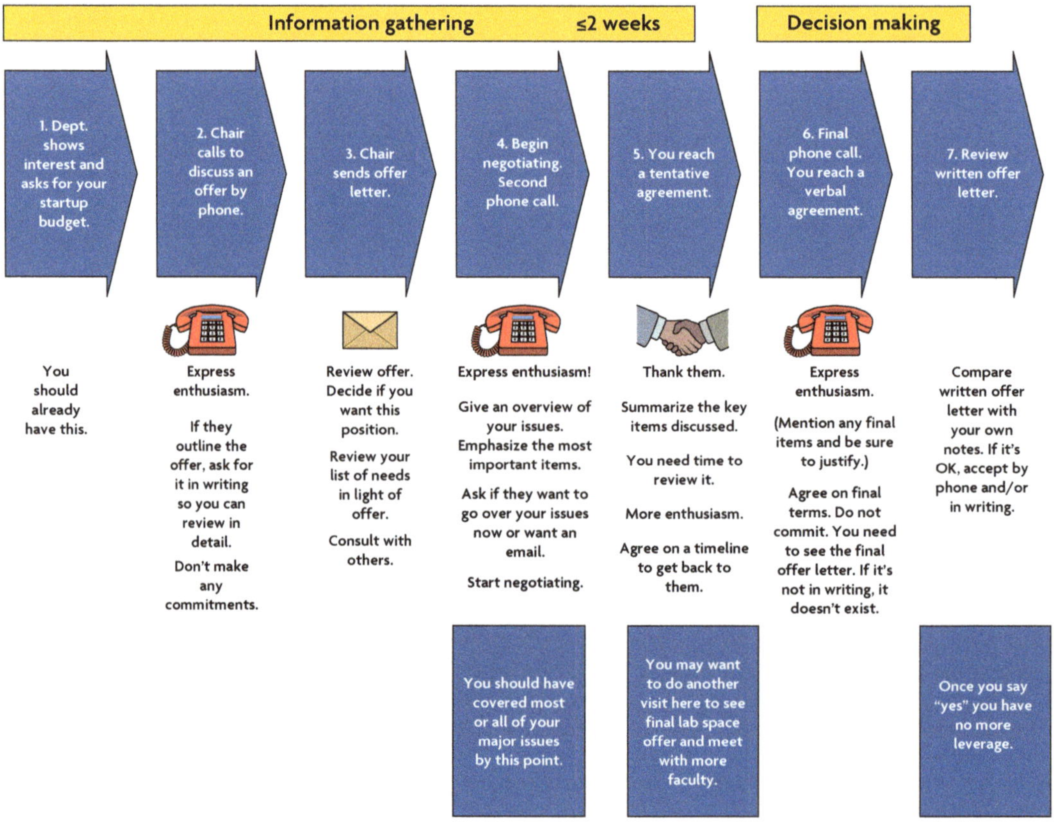

Beyond comparing the offer (Step 3) with your list of priorities, you should be reaching out to your trusted advisers to get their take. They may pick up things that you miss or be familiar with how this offer compares with others they have seen for candidates for similar positions.

In Step 4, you have identified areas that need addressing. You have emailed the list to them and schedule a call to discuss.

In Step 5, you have discussed and reconciled any differences and have the outline for the terms of a *tentative* agreement. You could ask for this in writing now. But you may also want one more opportunity to make sure that you have covered everything. One thing you may want to do is to conduct a second visit to take another look at lab space, bring your partner to look at housing options, and meet with additional faculty. Assuming no new issues pop up, after your visit you review the terms of the final agreement with them on the phone and then ask for it in writing. Assuming all is in order, you schedule a call to go over the final agreement (Step 6).

> *First, thanks for working with me on this offer. I really appreciate it. What you outlined sounds pretty much like what we've been discussing and I'm feeling very positive about this. Clearly, before I give you a definite "yes," I'm going to need to see all this in writing. If all is in order, then I think we'll have a deal.*

Ask them how they would like you to respond—by phone, in writing, etc.

In Step 7, you have carefully reviewed the final offer letter to make sure it reflects what you thought you agreed to. This may sound matter of fact, but you would be surprised by the frequency with which you will remember one thing, and they will remember something different. Provided the offer is what you agreed to, let them know by email or phone, sign the offer letter, and return a signed copy to them.

ALL THINGS ARE NEGOTIABLE

Embedded in the above sequence are numerous discrete negotiations about the specific terms of your agreement. The number of potentially negotiable items, listed next, is staggering, and hopefully you will have major differences over only a few of them.[71] The table below lists the most common negotiable items of an academic offer, separated into four categories: your salary, startup package, terms and conditions, and other factors to consider.

YOUR SALARY: IT'S A NUMBERS GAME

Salary is always the first thing on a job applicant's mind and is nearly always negotiable, up to a point. Note that we said "nearly always" because there are institutions or departments where the starting salary is nonnegotiable (often to mitigate the role of bias in the negotiation and offer process). These institutions will typically make this known during the recruitment process, and for many applicants it will be a relief to learn that they will not have to engage in what can be a stressful negotiation. For all other institutions the key is to divine the meaning of "up to a point." If you are lucky enough to have applied for a position for which the salary range was posted, some of the guesswork has been taken out of the process. Here is an example from a recent advertisement for an assistant professor.

> *The salary for this position is $141,714, which represents the base salary for this position, and is based on the university's good faith belief at the time of posting. The actual compensation offered to the selected candidate may vary and will ultimately depend on multiple factors, which may include the successful candidate's geographic location, skills, work experience, internal equity, market conditions, education/training, and other factors, as reasonably determined by the university.*

A salary of $141,714 is a comedically specific number arrived at in a way that exactly no one will be able to explain to you. Also, even though the posting says there is flexibility, no range is given. So, you still need to figure out where you are going to land. Let us use our fictional candidate Morgan Chen to walk through the following *step-by-step guide to finding the expected salary range.*

Primary negotiable elements of an academic job offer

1. Your salary	2. Startup package	3. Terms and conditions	4. Factors to consider (less negotiable or nonnegotiable)
• Amount • 12 months or 9 months? • Requirement for soft money contribution	*Recurring expenses:* • Staff salary • Supplies • Travel • Professional development assistance (funds for executive coaching or training) • Other (computing fees, shared facility/ core use fees) *One-time expenses:* • Lab equipment • Office furniture and computer equipment (on campus and/or at home) • Relocation expenses and/or housing purchase expense/ mortgage assistance including moving expenses, realtor's fees, temporary housing • Signing bonus	• Starting date • Lab/office space • Living expenses (institutional housing, housing allowance) • Startup funds' fungibility • Immigration and naturalization assistance (including fees and legal assistance in obtaining an H-1B visa and throughout the process of immigration) • Teaching responsibilities • Reduced or free service from campus facilities, such as machine or wood shops, instrumentation centers, such as NMR, etc. • Tenure decision timeline adjustments • Date by which candidate should respond to an offer • Time for candidate to resign from current position • Limits on consulting? • Clinical time (for MDs in a medical school setting) • Spousal job offer • For nontenure-track appointments: term of appointment; option to supplement income by teaching if needed	• Benefits: healthcare; dental; insurance; parental leave; spousal benefits; time off • Childcare: availability of childcare resources and referrals (also, care during research data collection or conferences) • Tuition benefits for children • Retirement benefits and 401K match

Step 1. Find the range. Unlike the advertisement above, most postings for academic positions will not include a salary range. In September 2025, there were 2,872 faculty positions in science and mathematics listed in the Chronicle of Higher Education, only ninety-three of which specified a salary range. Morgan could start his research on an expected salary with publicly available salary surveys.[48,49,50] The 2024 AAUP survey lists an average salary of $127,400 for an assistant professor (AP) at Cornell University across all disciplines.[50] Note that these numbers probably apply to similar peer R1 institutions. If Morgan were applying to an R2 institution or a small liberal arts college, the range he would find on Glassdoor or through word of mouth would certainly be lower.

But even at Cornell there is a wide range within the AP level. Glassdoor shows a range of $101,00 to $159,000 for new APs at Cornell.[72] What accounts for such a wide range? Most of it is because of academic discipline.

Step 2. Adjust for academic discipline. Various sources show that APs in science make on average about 25% more than all APs together.[73] Since the $127,000 salary for a Cornell AP cited was averaged across all disciplines, Morgan needs to increase his estimate by 25% to $159,000 because he is in science. But even within science there is a broad range, with some science disciplines paying considerably more than others.[74] The table below shows a sampling of first-year AP salaries in various sciences taken from a 2023–2024 survey of 642 academic institutions in the United States.[75] The position that Morgan is applying for is a combination of cellular biology and marine sciences, and the table shows that salaries in both disciplines fall about in the midpoint or average of the science salary range. So, no discipline-related adjustment to Morgan's estimate is needed. If on the other hand he was looking at an AP position in information science, which is at the high end of the pack, he would add another 20% to his expected salary.

Median salaries for first-year assistant professors in sciences at US academic institutions, 2023–2024

Discipline	Median salary
Agricultural science	$62,000
General biology	$70,000
Psychology	$71,000
Mathematics	$74,000
Chemistry	$75,000
Biochemistry	$82,000
Astronomy and astrophysics	$82,000
Earth sciences	$82,000
Physics	$83,000
Cellular biology	$88,000

(Continued)

Discipline	Median salary
Marine sciences	$90,000
Statistics	$93,000
Bioinformatics	$95,000
Computer science	$97,000
Information science	$105,000
Biomedical engineering	$106,000
Computer engineering	$110,000

Step 3. Adjust for the exact position. Morgan is trying to estimate his expected salary as a first-year AP. But the averages from most surveys include APs in that position for just one year as well as those in an AP position for two, three, or even more years who have had modest salary increases each year. So, he needs to pare back his estimate, and he guesses at a 10% cut (figuring that APs get a 3% per year salary increase for an average of three years, compounded) for a final estimate of $143,000.

Now Morgan is ready to set his ideal, happy, and bottom-line salary numbers. "I'd love $160,000, I'd be happy with $150,000, and there's no way I'm taking this job for less than $145,000." Morgan is ready to negotiate.

What's *your* number?

What if they ask you what salary you want? Some pundits will tell you never to answer this question, in the belief that making the first offer puts you in a weak position. If you ask for less than they are willing to pay, you have given up money. If you ask for way too much, you look like you have not done your homework or have unrealistic expectations. If by some miracle you ask for close to what they were going to offer you and they say yes, you may have once more left money on the table because you did not negotiate.

But what if, like Morgan, you have done your homework and have a 90% certainty that you know what the salary range for this job is? Now it seems like it could be OK to answer that question with "I'm thinking that $160,000 would be something I could live with for a nine-month salary. I'd also want to discuss some interim summer support, but we can get to that later."

Morgan has given them a mildly aspirational but reasonable number—in negotiation parlance, it is called an anchor. He knows it is probably high but not unrealistically so. The effect of an anchor is that it can make them reluctant to come back with a lowball number like $125,000 because it seems so much less than his number. Anchors can have a powerful effect on the other person's thinking[76]—even arbitrary numbers can fix their thinking for no rational reason.[77] But starting with a stratospherically high figure in the hope of anchoring the other side can backfire if it makes you seem either unaffordable or unrealistic. A possible compromise might be "Look, I'm not going to ask for a salary of $180,000 (an unrealistic anchor stated in a way that makes it clear that you are not asking for it)—I understand your constraints. But something in the range of $160,000 seems fair to me."

If you have no idea what a realistic range is and, for the reasons cited above, do not want to hazard a guess that could backfire, then say "I can't give you a specific number, but I would expect a salary that's in line with others with my experience in the same role." If this question comes very early in the interview process—for example, in a screening interview—try "I haven't really had a chance to research the salary range for this position, so I don't want to hazard a guess. I'm sure we'll be able to come to an agreement if you think I'm the right candidate." Of course, saying that in the moment in no way means that when they do make you an offer you are not going to negotiate it.

Start negotiating

Given the above, if their initial offer is $125,000, Morgan is definitely going to negotiate for more. He could start with "That sounds a little low to me. I'm looking for closer to $160,000," but a better approach might be "Based on my research, that looks like that's at the very low end of the range for this position. I'm hoping for something in the $160,000 range. My record, and the research program I'm building, justify my being at the higher end of your range. How can we make that work?"

As aptly pointed out by others, do not minimize the importance of seemingly small increments in starting salary.[37] The difference between a starting salary of $85,000 and one of $80,000, assuming a 3% salary bump each year and that you invested the difference conservatively, could net you an additional benefit of $150,000 over twenty years.

Summer at the beach?

Most academic salaries are based on a nine-month academic year (you get the summer off and no pay), which can be paid to you over twelve months either by default or by request. They are paying you for nine months of work, but no scientist in history has ever taken the summer off. You have three months in the summer when you will actually be working (in your lab, writing a grant, or on a field trip) and they are not paying you; it just looks like they are because they spread your paychecks over twelve months. Most institutions expect you to cover those months (or two of them) from a grant, but you do not have one yet because you are just starting. What to do? You ask them to cover your summer salary for your first two or three years. That becomes part of your negotiation and will get included in your startup package. Some institutions will do this (one report indicates that about one third of academic packages provided some level of summer support at startup) and others will not, but you will not know unless you ask and negotiate it.[71]

They giveth and they taketh away

Many institutions expect that at some point you will cover a percentage of your salary, maybe 30%–50%, from research grants, a policy called salary recovery. Sometimes the time for this to happen is defined (within five years, for example) or left unspecified. While this is not part of a salary negotiation per se, salary recovery options are definitely something to be considered when evaluating an offer, especially if you have competing offers and one requires a 10% recovery while the other is 40%. Other, more nuanced issues with salary recovery include using it to buy your way out of teaching or as justification for a raise.[78,79] Some institutions will tell you up front that if you bring in a certain dollar amount in grant funding, they will increase your salary by a predetermined amount.

In rare situations you may be able to turn salary recovery on its head to the benefit of your research lab. I know of a research hospital that recruited a new department chair to enhance its presence in a key area. The chair negotiated several full-time faculty salaries to be paid from hard money by the hospital. The chair also negotiated that should any of these hard money scientists receive grant funding that covered some percentage of their salary, the hospital would reallocate the equivalent dollar amount from its coffers to a research account that the funded researcher could use in an unrestricted manner for research costs. Sweet deal.

NEGOTIATING YOUR STARTUP PACKAGE: DO IT THIS WAY NOT THAT WAY

For a lab- or field-based research scientist, the amount of startup funds you get can make the difference between your first three years being spent doing well-funded productive work or begging for freezer space from the lab down the hall. The way to develop and justify your startup budget is a bit different from how you would develop a budget for a research grant.

A. Define what they expect of you: Scientific productivity and future research grants.

B. Outline the big research objectives you will pursue to accomplish what they expect of you in A.

C. Develop a detailed research plan that will enable you to achieve the big objectives in B.

D. Create a budget that will allow you to successfully undertake the research plan in C.

This means that the amount of money you ask for in part D is directly tied to and justified by the plan in C and the big objectives in B.

Justify your requests

Every job offer negotiation is based on this: *You both want YOU to be successful and productive. This is one of your and their underlying interests.*
• What level of productivity does the institution expect of you? • What do you need to attain this level of productivity? • If you do not get some of the items on your list, how does that impact your productivity?

In this approach, you *start* with what they expect from you (create a successful research program and get external funding). In a grant application that you may be more familiar with, you start with your research objectives (see figure below). If they offer less than you think you need, the top-down approach enables you to negotiate more *not because you just want more money but because you and they want you to succeed.*

Top-down vs. bottom-up justification

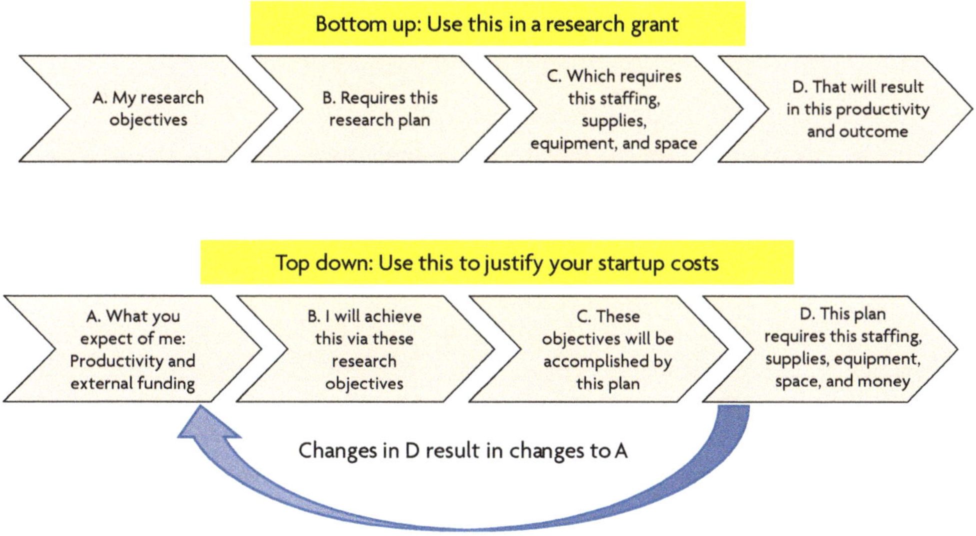

SAMPLE DIALOGUE

You have sent them your detailed startup budget as they requested; it came to a grand total of $2,500,000 over three years, not including your salary. The dean wants to set up a Zoom call with you to discuss it before she puts numbers in an offer letter.

Dean Halliday: "Thanks for sending us your startup budget request. We want to support you and your program as best we can because your success is important to us. I'm sure you realize that these are difficult times for universities and we're doing our best to balance all the conflicting needs we have. I think we can come close to what you asked for in your budget. We'd love you to come here and we've been able to put together an offer that will give you $1,700,000 over three years. I need to tell you that this is one of the biggest startup packages we've ever done, so it represents a serious commitment for us and reflects our confidence that you are the right person for this position."

You: "Wow—thank you so much, Dean Halliday. I really appreciate your confidence in my program and that your team thinks that I'm the right person for this role. I have to tell you that as I've looked at my options over the past months, everything I've seen at your institution tells me that this would be the right move for me. So, thanks so much for going over my budget list. At the same time, I do hope we can get closer to what's in that budget. As I'm sure you know, I put it together with the sole objective of ensuring that by the end of the three-year startup period I will have either received or will be about to receive significant external support. For that to happen in this incredibly competitive climate I need to be very productive; and to be *productive I need to have my lab staffed and running as quickly as possible. I honestly believe that the budget I sent gets us there. I don't think that $1.7 million will allow us to do that, and as I look at the budget, I wouldn't really know what to reduce without cutting the legs from under the program. So, yes, I'm extremely appreciative of your offer, and yes, I am hoping that the $2.5 million number is where we'll end up."*

Dean Halliday: "OK, well, that could be a problem for us. I'll tell you what, let me go back to my team and see what we can do to get closer to your number. While we're doing that, there were a few other items I wanted to discuss with you that will be in the letter ..."

> *You need to adopt the mindset that you are not* asking them for money. *They are* investing in you *to accomplish the goals that the two of you agree are essential for the success of your program.*

The time to start putting your startup package together is before you send in your first job application (see the figure at the beginning of the chapter) so that you have a good idea of what you will need before you talk to anyone. Knowing your startup needs can help you screen possible job options. If you find out via the grapevine or in early interviews that a particular institution has never given anyone even 50% of what you think you will need, you can back out early or start anchoring them on your needs so they can decide if you are a viable candidate sooner rather than later. Note that the type of institution you are applying to also needs to be considered. If you are applying to an R2 or teaching institution, you should not expect the same level of startup support as you might get at an R1 institution.

Example of a three-year startup budget

Three-year budget				
Personnel*	**Year 1**	**Year 2**	**Year 3**	**3-year total**
Postdoc	$35,000	$72,000	$74,160	$181,160
Grad student		$50,000	$51,500	$101,500
Grad student			$50,000	$50,000
Research assistant	$30,000	$60,000	$61,800	$151,800
Lab manager	$60,000	$61,800	$63,654	$185,454
Sub total	*$125,000*	*$243,800*	*$301,114*	*$669,914*
Fringe benefits 30%	$37,500	$73,140	$90,334	$200,974
Personnel total	**$162,500**	**$316,940**	**$391,448**	**$870,888**

(Continued)

Three-year budget (cont.)				
Other	Year 1	Year 2	Year 3	3-year total
Equipment (see attached list)	$755,000	$150,000	$50,000	$955,000
Consumables (see attached list)	$30,000	$50,000	$51,500	$131,500
Publication fees	$4,000	$8,000	$12,000	$24,000
Core facility use	$10,000	$20,000	$30,000	$60,000
Professional development	$15,000	$15,450	$15,914	$46,364
Travel and meetings	$15,000	$25,000	$30,000	$70,000
Misc.	$15,000	$15,450	$15,914	$46,364
Other total	**$844,000**	**$283,900**	**$205,328**	**$1,333,228**
Grand total	**$1,006,500**	**$600,840**	**$596,776**	**$2,204,116**
* Line items increase by 3% per year to account for inflation, except where they are phased in over time.				

Your startup funds are just one part of your startup package

The above is an example of a startup budget for a lab-based scientist. There will be dramatic differences in such a budget between disciplines. An assistant professor of mathematics or theoretical physics would have very different needs than a high energy physicist or a molecular biologist. In our budget we are using figures that would be typical in 2025 for biomedical science. This is a high-level budget in that it does not show the detail breakdown within the equipment and consumables categories. These would be put in separate attached spreadsheets so as not to obscure the big picture.

Phase in your expenses

In some categories of the budget the costs have been phased in over three years. You will not be hiring all your staff or spending at your full rate on the day you start work. Settling in, recruiting, and hiring staff takes time, especially if you are trying to set up a lab (which, no matter what anyone tells you, will not be ready until six months later than they said it would) and an office, and moving into your new house. So, for most categories, do not budget for a full year of expenses in your first year. That will yield budget savings that you can apply to other categories.

Major categories

The number of items that can be included in your startup package can exceed twenty,[80] but the major categories in terms of dollar value (not necessarily in order) are the following:

- Personnel (salary)
- Major equipment
- Small equipment
- Supplies and consumables
- Everything else

Personnel

What do you need in your first three years to get your research up and running? A typical package for a small lab might include a postdoc, a technician, a lab manager, and two graduate students. Seek guidance from the offering institution about their graduate student and postdoc salaries. Make sure that you ask about their fringe benefit rate—a postdoc getting a salary of $70,000 would also get benefits (health insurance, retirement plan, social security, etc.) of another 25%–35%. You can either spell that out as a line item or total it and call it "salary plus fringe."

The budget above includes only a partial year salary for the postdoc and research assistant in the first year for the reasons stated above. An exception to this is the lab manager, who is budgeted for early in year 1. If you are starting a research lab, getting a good lab manager on board quickly will be invaluable in helping you get your lab set up, equipment and supplies purchased, lab policies and procedures in place, etc. Once you have done that you can bring in your first postdoc and a research assistant about halfway or more into your first year, but even that could be aggressive. The first grad student is not added until year 2 because grad students need a lot of time and attention when they first start, and you will not be able to give them either in your first year. Also, grad students often do not start in a lab until they have done a brief rotation in that lab. That gives them and you time to decide if there is a good fit. Take your time accepting your first student—they are going to be with you for years.

Your budget also needs to account for inflation. The budget shown uses a 3% inflation factor, which was obtained using the calculator on the US Bureau of Labor Statistics website comparing the buying power of $1 in 2024 and 2025.[81]

Equipment: It's all in the details

Major equipment. The definition of major equipment will vary between institutions, but typically anything costing more than $5,000 should be listed here and specified in as much detail as possible. Showing them a list with each manufacturer's name, model number, and cost (Thermo Scientific™ GENESYS™ 180 UV-Vis Spectrophotometer, $10,800) is more convincing than listing it as "UV-Vis spectrometer."

Small equipment. You can decide what qualifies as small equipment, for example, anything costing more than $100, but you should specify that. List items in as much detail as you have time for, with manufacturer, model name, and price. Unlike with personnel and consumables, where you add more people and costs over time, do the opposite with equipment. You will likely be purchasing most of your equipment in year 1 and less in the subsequent years.

Supplies and consumables: Do I need to list every Sharpie?

There is a school of thought that suggests that more detail is better than less. It cannot hurt if you have a separate supplies and consumables budget page that lists up to fifty items, their unit cost, the number needed per year, etc. This lends credibility to your numbers. Phase in your consumable costs over the three years to reflect that you are starting slowly in year 1 and ramping up to year 3 when you are fully staffed.

Everything else

Note the line item labeled "Professional development." This refers to things like executive coaching, training workshops, or programs in management and leadership. More and more academic and industry organizations are realizing that scientists coming from an academic background (as nearly all do) have little or no training in managing other scientists, being an effective mentor, leading a team, or negotiating conflicts. Including this line item in your budget shows that you are one step ahead of them and that you will be proactive in acquiring these skills.

Also included here is "signing bonus." Signing bonuses are nearly universal in industry for mid- to senior-level positions and are becoming more common even for junior positions, especially when specific or highly valued skills are required. Until relatively recently, signing bonuses were almost nonexistent in academia but are becoming more and more available, especially in cases where a highly qualified or accomplished individual is being recruited and in cases where a candidate has multiple offers to choose from. In the latter case, when the institution has gone as far as it feels comfortable going in salary, a one-time signing bonus can make the difference. In other cases, signing bonuses may be lumped into a very generous relocation expense so the institution can get away with not calling it a signing bonus.

TERMS AND CONDITIONS

The previous items on your list are principally monetary in nature. The next set of items relates to various concessions and commitments they will make.

Lab and office space

Along with salary and startup funds, lab space is one of the three big-ticket items in your package. Once you negotiate your space, that is what you will be working with, at least until you start bringing in grant money. Think carefully before agreeing to anything.

Be especially cautious with space that needs to be renovated, which almost all lab space does, especially if you have unique needs. Lab renovations take approximately forever, so try to negotiate into your agreement some date by which your space will be ready; if that is not possible, have them specify an interim space to use until renovations are complete. Try to get them to be as specific as possible and tell them you would like it all in the offer letter, including the name of the building, the room numbers, number of square feet, lab furnishings, etc.[82,83] Better to be very specific up front than disappointed later.

Renovations to existing space are expensive, so you do not want to have these costs included in your startup budget. They are providing you with space, and it is their responsibility to ensure that it is up to standards and meets your needs.

Teaching responsibilities

Most academic departments have boilerplate expectations for faculty teaching responsibilities that are almost always negotiable to a degree that depends on the type of institution it is. Primary teaching institutions will want you to do more teaching than R1 institutions. Teaching loads are designated by the abbreviation (X,Y), where X is the number of courses taught in the first semester of a given year and Y is the second semester. For a new hire at an R1 institution, the teaching requirement might be no teaching in your first year (0,0), one course in the second semester of the second year (0,1), and one course per semester thereafter (1,1). For more teaching-oriented institutions, the second and third years could be anywhere from (2,2) to (4,4), the last case being one where doing research will be challenging. As with your startup package, you will be negotiating your teaching load based on the shared goal of starting your research program and getting funded quickly.

> *I'm concerned I won't be able balance my schedule if I have too heavy a teaching commitment early on. If you can keep my teaching to one course in my first year, I'll be more confident in my teaching quality, and I will be able to finish several manuscripts that will support my grant application.*

Another negotiable element is what is referred to as "course preps." These are courses you have not taught before and for which you will have to create your own material for a one-semester course from scratch. This is a lot of work and eats significantly into your research time. You may be able to negotiate that you will have only one new course prep in your first two or three years.

Finally, it is common practice that you can "buy out" some or all of your teaching if you bring in grant funding that includes some salary for you. Your institution will use the salary line item in your grant (which they are already committed to paying) to pay someone else to cover your teaching obligations. Policies on this practice vary between institutions.

Justifying "quality of life" items

• Benefits (health care; dental; insurance; parental leave; spousal benefits; time off) • Childcare: availability of resources • Teaching load and committee responsibilities • Sabbatical • Spousal job opportunities • Remote working
Your underlying interest: "If I have just a small amount of extra free time, I can get my grant/manuscript/new course syllabus done sooner. I need to have a healthy work/life balance so I can be productive." **Their underlying interest:** Legitimate and/or pressing teaching needs or equity with others.

Starting date

In academia, this is almost always negotiable unless you are being hired by an institution that desperately needs you to teach Introduction to String Theory starting on September 14 and they specified that well in advance. But even then, they may grant you flexibility if they really want you. If this is an important item for you (which you have signaled soon after they made you an offer, as opposed to at the last minute), it is in their interest to find a solution. Many postdocs, for example, want to finish an important project in their current lab to put them into a better position to get a high-profile publication and increase their chances of funding when they take a faculty job. In that case, a delayed start is in the interest of both the postdoc and the hiring department. A delay of months is relatively easy but even a delay of a year is not unheard of.

Startup funds' longevity and fungibility

How long will your startup funds be available to you? *You* want these to be available for as long as possible, ideally forever. *They* may want you to spend them in three or five years. This is negotiable, and you should push for the longest possible time, including for as long as you are there.

If your startup funds are granted as a lump sum, you can presumably spend them on whatever legitimate costs you want. If on the other hand the funds were divided into budget categories (e.g., personnel, equipment, etc.), then you need to know whether funds in one category can be used for another (called fungibility) if needed. If they budget by category and tell you that funds are not transferable between categories, you should negotiate that because there is usually no good justification for it. If they tell you that this is just "their policy," remember that policies are made by administrators and can therefore be changed by administrators. Categorized budgets can be a real constraint, especially if you are new to the budgeting game and some of your estimates are off, or if you are having a hard time filling a position and you have unexpended salary funds you would like to use for other things.

Tenure decision timeline adjustments

Find out when the tenure decision clock will start for you. Institutions have differing policies on how long after you are hired you need to go up for your tenure decision, but six years is common. If you were hired very late in the academic year, you can make the case that your six-year clock should not start counting down until the next academic year.

Dual career couples: Just when you thought it couldn't get more complicated

While finding and negotiating for one position is complicated, doing this for two different positions simultaneously is even more so. Many scientists have partners who are also scientists. A 1998 study suggested that 89% of female physicists were married to scientists, and 45% were married to other physicists.[84] A 2008 study from Stanford University found 36% of full-time faculty in their study were married to other academics.[85] The desire of a married or partnered couple to remain together introduces additional complexity and emotional stress into the search process.[86]

The issues and decisions involved in negotiating a dual hire are so different from couple to couple that generic guidance is not possible. Nonetheless, the following outlines some of the strategic decisions you will need to make and factors to consider if finding a position for your partner is an important consideration.

First, you need to decide at what point in the interview or negotiation process you should reveal your desire for a partner hire to the hiring institution. The case for doing so early in the process is that it gives institutional officials plenty of time to explore options for your partner. This is a good strategy with institutions that have a stated policy of accommodating spousal or partner hires.[87] Some institutions post such policies on their websites,[88] while others can be found in the "Partner Hire Scorecard,"[89] a downloadable list of institutions, organized by name and their likelihood of accommodating a partner hire.

For institutions without such stated policies, revealing that you are looking for a partner hire too early in the process may dampen their enthusiasm before you can impress them with your job talk and winning personality.

When in the recruitment process should you reveal you are looking for a partner hire?

Sooner		Later	
Pro	***Con***	***Pro***	***Con***
Gives them time to find a solution. Choose this if they have an explicit dual career policy.	They may rule you out early in the process.	They have invested time in you and possibly made an offer. They do not want to start over again if you turn them down.	They may feel misled by you for not telling them sooner; if no second position is possible, you have wasted a lot of time.

But waiting to let them know until later has its own complexities. On the positive side is the psychology of "sunk costs." The more time they have invested in your candidacy and in crafting an offer, the more motivated they may be to find your partner a position. If you decline their offer because they cannot find a position for your partner, they will have to start all over again with a new candidate. On the negative side, they may see this tactic for what it is—manipulative. From your perspective, if you wait until late in the game to reveal your need for a second position and it turns out to not be possible, you have wasted a lot of your time that could have been better spent on other opportunities.

You and your partner need to decide how important having a position for your partner will be to your decision-making process. If you get offered a job, but there is no prospect of your partner getting an offer, would you turn down the offer? As with the timing decision, the decision to reveal whether a second position is a deal-breaker is a complex one. If you say that it is a deal-breaker early in the process, before they have made a commitment to you, it may be easier for them to rule you out. If you say it is a deal-breaker late in the game, they may feel annoyed and manipulated whether or not they can find a position for your partner.

What if a position for your partner is something you would like to have but is not a deal-breaker? By letting them know that you will consider the job even if they do not make your partner an offer, you are lowering their incentive for finding your partner a position. However, if you are an exceptionally strong candidate, they may use an offer to your partner as a way to incentivize you, even if it is not a deal-breaker.

While this process is complex, it can be managed if you and your partner work as a team and evaluate your options as the recruitment and negotiation process progresses. *Deciding what is most important for you and your partner, both professionally and personally, is the most important thing you can do to prepare for this process.* Regardless of what you think when you begin, what either or both of you will accept as an outcome will *likely change* during the search process, especially if it is taking too long, jobs are limited, or unexpected opportunities appear. It is almost inevitable that one of you will take a position that is less than ideal, in order for the other to take one that is ideal or nearly so.[85,90] A decision like this is never easy and can stress your relationship, especially if the process seems to value one person's career more than that of the other. In times such as this, your commitment to the relationship and your joint ability to engage in difficult conversations will work in your favor. Beyond that, as in all other aspects of your job search, research and data are your best friends. Find out what, if any, policies the institution has on spousal or partner hires and try to find out whether they have made such hires in the recent past and what the outcomes were.

NONTENURE-TRACK ACADEMIC RESEARCH POSITIONS

As mentioned in the introduction, the percentage of nontenure-track academic positions has increased dramatically over the past twenty years.[9] In 2023, nearly 70% of the US academic faculty workforce were nontenure-track full- or part-time employees.[91]

If you are applying for such a position, your negotiating power will be limited but not zero. Research faculty who bring their own research and salary funds are in the strongest positions because their grants come with overhead funding that goes directly to the institution to support the administration and infrastructure associated with laboratory or clinical research.

From the institution's perspective, externally funded research faculty are a very good deal. The institution did not have to go through a time-consuming search process to find you, you are not taking up a tenure-track slot, you are paying rent, and the institution has fewer obligations to you. Because of this, you are in a strong position to negotiate many of the same items that we covered for tenure-track positions.[92] While research faculty get paid through the institution (from the funds their grants bring in), they are paid on a separate scale than tenure-track faculty and may or may not have access to employment benefits.[93] It is worth noting that nontenure-track faculty pay and benefits are typically considerably better in institutions in which they are represented by a union than in those without one.[94,95]

There is still room for negotiation, especially if you know what others in similar roles get paid. Items that may enter into your negotiation include the following:

1. The length or term of your appointment (longer is better).
2. Fees: The institution may ask you to pay for use of shared services like computing and core laboratories (you should argue that these should be covered by your grant's overhead payments).
3. The option to take on teaching responsibilities to supplement your income or funding should you experience a gap in grant funding.
4. Their willingness to provide research bridge funding for some predetermined time should you experience a funding gap. Pay special attention to how the start and end dates of your appointment do or do not line up with your funding cycle.

CHAPTER 10 SUMMARY

Follow the typical academic negotiation sequence

- Expect calls requesting your startup budget, then a preliminary offer, followed by back-and-forth negotiations, and finally a written offer. Never say "yes" until you see everything in writing and have time to review with advisers.

Research salary ranges thoroughly using multiple sources

- Use AAUP surveys, Glassdoor, public salary databases, and discipline-specific data. Adjust for your field (sciences pay ~25% more than average), institution level, and experience. Set three targets: ideal, acceptable, and bottom-line numbers.

Negotiate startup packages using a top-down approach

- Start with what they expect from you (successful research program and funding), define your research objectives, create a detailed research plan, then build a budget that supports that plan. This justifies every dollar as necessary for mutual success.

Create detailed, phased-in budgets with specific equipment lists

- Phase in expenses over three years (you will not spend at the full rate immediately), include manufacturers' names and model numbers for equipment, account for 3% annual inflation, and add professional development funds to show you are thinking strategically.

Many nonsalary elements are negotiable

- Include teaching load (especially course preps), lab/office space specifications, startup fund flexibility and timeline, starting date, tenure clock adjustments, and terms like fungibility between budget categories. Get specific details in writing.

Nontenure-track research faculty have significant leverage

- If you bring external funding, you are valuable because you pay overhead while requiring minimal institutional investment. Negotiate appointment length, fee waivers for shared services, teaching options for income gaps, and bridge funding commitments.

11 Simpler but Not Easier

The Strategic Guide to Industry Job Negotiations

- **Industry job offer sequence**
- **Industry job offer elements**
  - Base salary
  - Signing bonus
  - Annual bonus
  - Stock options
  - Relocation expenses
  - Start date
- **Chapter 11 summary**

INDUSTRY JOB OFFER SEQUENCE

With some exceptions, all of what was said about job offer negotiations, especially the negotiation principles in Chapter 9, applies to industry. Here we will highlight areas and practices that are unique to industry.

With regard to complexity, comparing a job offer negotiation from industry to one from academia is like comparing checkers with chess. Except for industry jobs at the highest levels, there are far fewer options and moving pieces. While that may make the calculus of an offer simpler, it does not necessarily mean that the process itself will be any easier.

Industry job offer negotiation sequence

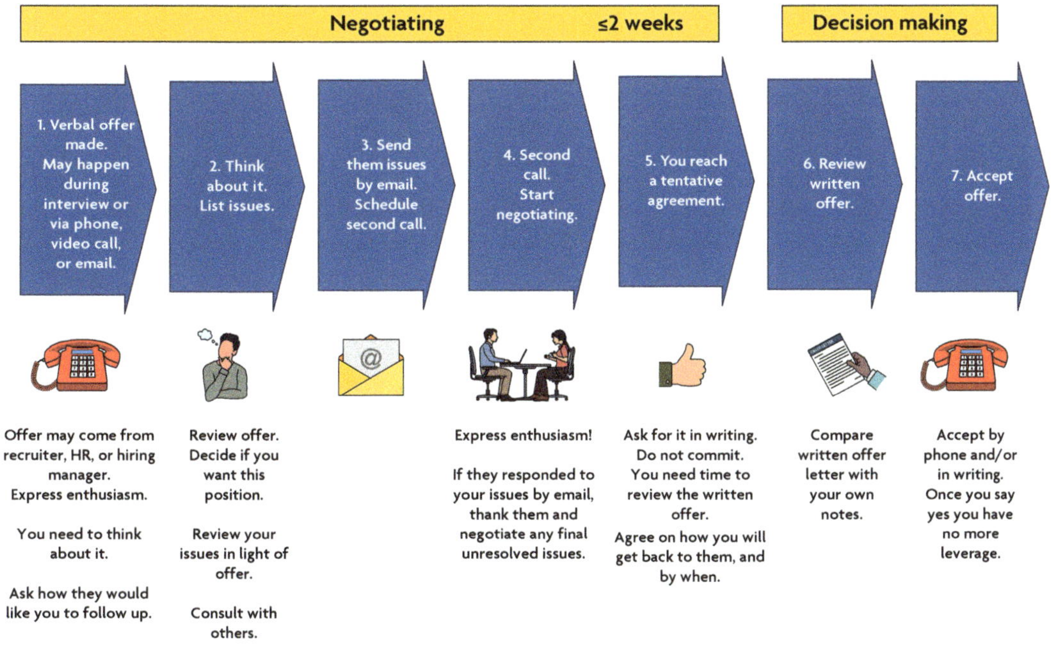

The figure above shows a typical offer sequence for an industry job. In many cases your first contact will be from either their HR group or, for more senior roles, a recruiter or headhunter. It is worth noting that in times of hot industry job markets you may get contacted by a recruiter without having applied for any job. Recruiters have stealthy ways of getting your contact information, including from your friends. If you do get contacted by a recruiter, the first thing you should do is to find out whether they have been retained to fill a specific position and, if so, what it is. Some recruiters work on speculation or "spec," meaning they try to round up a batch of résumés from people like you and then shop them around to various companies. You need to decide if you want them to do this with *your* résumé. Such résumé shopping can backfire if your current employer finds out that you are actively looking for another job.

The recruiter or HR staff member will usually call you to set up an initial screening interview either on video or by phone. Everything in Chapters 3 and 4 about preparing for such interviews in academia holds here. You should have done your homework on the company and have intelligent questions for the interviewer. Take the screening interview seriously, since if you fail you will go no further in the process.

INDUSTRY JOB OFFER ELEMENTS

The major elements in an industry job offer include the following:

- Base salary
- Signing bonus
- Annual bonus
- Stock options
- Relocation expenses
- Start date

Let us take them one by one.

Base salary

In many cases, including nearly all entry-level jobs in industry, this is the only thing that is negotiable, and you should negotiate it. For junior or entry-level technical positions, you are not going to have a lot of leverage unless you have good justification. Justifications include expertise in a valued or unique technical skill or having a master's degree. If you are a Ph.D. applying for a scientist or supervisory position, evidence of prior managerial or supervisory experience will be a plus.

Finding out what the salary range is (especially the upper bound) will require research since many position postings in industry do not include it. Some states have enacted or are enacting laws that require them to do so now or in the future, so this may change. Some companies may tell you the salary range early in the game to make sure that it is acceptable to you. As with an academic job, posted ranges are often so broad as to be uninformative. Some of the online databases listed in Chapter 9 can be useful for finding salaries at specific companies. Here again, though, confusion may reign. In academia there are well-defined job titles—an assistant professor at Stanford means pretty much the same thing as one at Columbia. In industry it is not clear how to compare "Research assistant 1" at Biogen with "Lab assistant 2" at Genentech. Nonetheless, some information is better than no information, and places like Glassdoor,[54] H1B Database,[51] Indeed,[52] and online bulletin boards may be your best option for calibrating your thinking.

The more research you have done and the more you know about what others in similar positions get paid, the better you will be at gauging how much to ask for. If they are offering a starting base salary of $95,000, your response could be "I was hoping for something more in the range of $120,000." If they ask where you got that number, the answer is "I've looked online at various comparable jobs in peer companies and have spoken to friends employed at similar places" or that you have another offer at that salary.

Factors that can be used to negotiate your salary in industry

Technical or support positions (non-Ph.D.)	• Master's degree • Technical skill in a valued technique • Technical skill with complex instruments • More years of prior experience than most • You have another offer
Ph.D.-level science or supervisory positions	• Evidence (especially publications) of expertise in a scientific area that they need • Documented managerial or supervisory skills • Relevant experience in a similar role at another company • You have another offer

While there are many reasons you can muster to justify a better salary, there are also reasons you either should not use or are not convincing. These include the following:

- Personal life circumstances and needs. The fact that you may have a high financial burden, for personal, family, or any reason, is not a justification for a higher salary.
- Your previous salary was higher than they are offering you. There is no reason not to bring this up, but most employers will not see this *on its own* as a compelling reason to pay you more. However, in conjunction with other factors, such as if they really want you or if you can marshal some of the reasons listed above to justify why you are worth more, it could help.

Signing bonus

A signing bonus is a payment that companies provide senior and some mid-level employees as an incentive to take a job. In industry, signing bonuses are common, even for entry-level positions. While a signing bonus may not be offered in your initial negotiations, it may be added as you get closer to an agreement for the following reasons:

- You have rare or valuable skills that they need.
- They are under time pressure to fill the position.
- They want you, and you do not seem as enthusiastic about the job as they would like you to be.
- You have another, possibly more attractive, offer and they want to sweeten theirs.
- You want a higher base salary than they want to pay, and you or they suggest that a signing bonus could compensate for that.

If you have more than one job offer and they are essentially equivalent, a signing bonus could be the decider. It is not recommended that you reveal that the bonus is what will tip the scale for you, since that makes it look like you are less interested in the company per se than the money. On the other hand, there is nothing wrong with asking whether there is a signing bonus for this job. If there is not, and they really want you, having asked could signal to them that this could help you decide.

Annual bonus

Except for the leanest startup, nearly all science and technology companies offer some sort of bonus system. One source indicates that 74% of IT companies offer bonuses of one type or another.[96] Bonuses are most typically performance- or incentive-based, meaning a reward or promise of a reward for achieving a company goal. Bonuses are typically tied on a percentage basis to both company and group performance over the past year, so they reflect both corporate and team achievements. In lean times there may be no bonuses, while in rich times bonuses can be prevalent and generous—anywhere from 3% to 30% of the employee's base salary, depending on their position. While bonuses are most often allocated by a formula, that does not mean they are non-negotiable. An offer letter will typically specify something like "Your base salary will be X, and you will be eligible for a performance-based bonus of up to Y%." So, what you will be negotiating is the maximum possible bonus you could get. While lower-level employees may be eligible for bonuses, it would be unusual for the amount to be negotiable until you reach middle management levels.

Stock options

Stock options are very common, especially in early-stage pre-IPO companies. When you join a company that has not yet gone public (the company's shares are not traded on a public platform like NASDAQ), stock options are a mechanism by which you can buy a certain number of company shares at the pre-IPO price. If and when your company goes public, you can sell your shares at, hopefully, the higher post-IPO price. We are not going to go into the complexities of options here except to note the obvious—if you believe in the company, stock options are a real incentive, and negotiating for more when you are hired can pay off when the company is successful.

Stock options are also common for established public companies, where you may be granted the ability to buy a certain number of shares upon joining and annually thereafter. If the company's share price increases, you can sell yours at a profit or hold them in anticipation of bigger gains in the future.

Relocation expenses

If you need to move your home and/or family to take an industry job, they will offer you relocation funds, and with most organizations you will not even need to ask. The amount they offer may be negotiable, especially if you have unique or special circumstances, e.g., the move is coast to coast, you are moving from a low cost of living place to one that is higher, or you have a large family or one with special needs (such as a disability). If you are going to ask for more, be sure to have estimates of costs. Also included in relocation expenses could be the cost of temporary housing while you search for something permanent and the cost for you and a partner to make one or two visits to search for suitable housing.

Start date

Unlike in academia, significant delays in starting date are uncommon in industry. Especially for entry-level positions, they may want you to start instantly. Companies have a financial interest in getting positions filled as quickly as possible to enhance or maintain productivity. For more senior positions there will be more flexibility; more than a three-month delay would be unusual.

CHAPTER 11 SUMMARY

Industry negotiations can seem simpler than academic ones but still require a strategic approach

- Fewer elements are negotiable compared to academia (no startup packages or teaching loads), but all the same negotiation tools and principles apply. Entry-level positions may only have the salary negotiable, while senior roles have more flexibility.

Research salary ranges and prepare justifications for requests

- Use Glassdoor, H1B Database, and Indeed since industry job titles vary widely between companies. Justify salary increases with factors like advanced degrees, specialized technical skills, management experience, or competing offers.

Understand the typical industry sequence and recruiter dynamics

- Initial contact often comes from HR or recruiters. Verify if recruiters represent specific positions or are shopping résumés on spec, as the latter can backfire if your current employer discovers you are job hunting.

Signing bonuses are common and strategic negotiating tools

- Unlike academia, industry frequently offers signing bonuses even for lower-level positions. They are used when companies need specific skills, face time pressure, want to increase your enthusiasm, or cannot or will not increase base salary.

Annual bonuses are performance-based and negotiable at higher levels

- Most tech companies offer bonuses ranging from 3%–30% of base salary tied to company and individual performance. While formulas determine amounts, the maximum percentage can be negotiated, especially for management positions.

Stock options can be valuable, especially at pre-IPO companies

- Options allow you to buy shares at predetermined prices. For pre-IPO companies, this could mean significant gains if the company goes public. Even at public companies, options count on future stock price appreciation, so it is worth negotiating for more shares.

12 From Scientist to Science Leader

Landing Management and Leadership Positions

I am skilled at envisioning opportunities where others see problems, building alliances where none seem possible, and resolving conflicts when no agreement seems evident. I am able to bring organization and focus to scientific endeavors that by their nature are unpredictable.

—From profile of applicant for a senior leadership position

(Continued)

This and the following chapter cover applying for jobs that are as much or more management and leadership than science but that nonetheless require a scientific or technical background or expertise.

In what follows, I will be using the terms *management* and *leadership* together or interchangeably because I believe that in science the two are inextricably linked. It has been said that good managers help people do things right and good leaders inspire them to do the right things. In my view, good managers need at times to inspire just as much as good leaders need at times to instruct. The combination of both characteristics enables the best managers and leaders to create high-performing scientific organizations.

We are not all cut out to be managers or leaders. As discussed in Chapter 1, many of us love doing science and wish we could just keep doing it and not have to deal with the interpersonal issues that come with managing others. But we also want to advance in our careers, get promoted, and be financially secure. The way nearly all science-based organizations are structured makes it difficult to achieve those things without taking on some managerial or leadership responsibilities.

So, if you want to advance you have two options. The first is to put your aversion to managing aside, move up the managerial ladder, and hope for the best. The second is to put your aversion aside and at the same time *learn and acquire the skills you will need to succeed in your new role*. This chapter provides guidance for the second.

- **Appendix 3. Management/leadership job description**
- **Appendix 4. Cover letter for management/leadership candidate**
- **Appendix 5. Sample management/leadership interview questions**

DECIPHERING POSTINGS FOR SCIENCE MANAGEMENT AND LEADERSHIP POSITIONS

Don't be misled by this

If you are thinking of applying for a management or leadership job role, the first thing you need to know is what kinds of skills and experience they are looking for. For this you need to decipher the position descriptions in job advertisements, just as we did for scientist positions in Chapter 3. Before we go into postings in detail, let us start with a high-level examination of job postings for management positions to see what organizations say they are looking for.

Appendix 1 shows excerpts from five postings found online for scientific managerial and leadership positions. To help us decipher these postings I divided the qualifications for each position into three categories (the postings were not written this way):

1. Science-related
2. Managerial/leadership/interpersonal-related
3. A combination of both

This sample of position qualifications, while small, illustrates that employers seeking to fill leadership or management roles seem to have dramatically different requirements. They may list mostly or all science-related qualifications, as in position #1, mostly managerial and leadership qualifications, as in position #5, or a combination, as in the other three (see figure below).

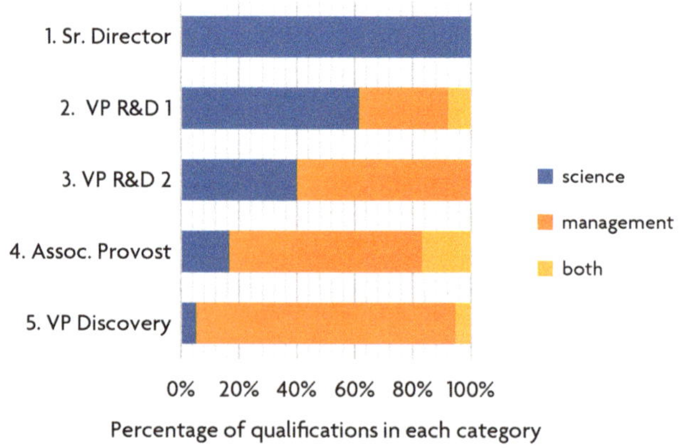

Despite the fact that positions #1 and #2 do not specify much or any managerial skills needed, *do not be misled by this*. VP of Research and Senior Director positions will require *significant* managerial and interpersonal skills. The fact that these descriptions do not mention this, especially the Senior Director role, which lists *none*, could mean any number of things, including that the position description was written by a scientist. Do not assume that because the position requirements do not include managerial skills that they are not a key part of the job or that you will not get asked questions about your managerial skills during an interview. *Assume the opposite*. This applies to every science management/leadership posting you will encounter. In the following you will see how to use this insight when applying for a job.

Reading between the lines

To delve deeper into a management/leadership job posting, we are going to introduce a new character, Sam Haberman, whose résumé is in Appendix 2. Sam has followed the instructions for preparing a résumé for industry we gave in Chapter 3, including using strong action phrases ("planned and created," "designed and implemented," and "built") to characterize his experience.

Sam is an experienced industry veteran who will apply for the VP R&D 1 position from Appendix 1. This position description is more heavily weighted in scientific qualifications over managerial and interpersonal ones, but Sam will not be misled by this.

The full posting for this job is in Appendix 3. The text of the posting is on the left and the interpretation of what each element of the posting means is on the right. While this position description may look daunting, it is simpler than it looks. The majority of the elements listed are "core" qualifications, which include basic scientific skills, maintenance of professional standards, and excellent communication capabilities. Beyond including some key words referencing these skills in an application letter or résumé (to ensure he does not get rejected by the AI reader), he will only need to address these if asked.

The interpretative comments to the right will guide Sam's understanding of the position requirements and how or whether he needs to address them in his application letter or elsewhere. When Sam compares his résumé with the position requirements, he will decide that while he is a good candidate for the job, he needs to emphasize some features of his background and experience to make that case. The items that Sam actually needs to address in the position description are underlined and numbered. These include interpersonal and managerial skills as well as specific scientific or related skills that a typical candidate might not have.

This position is a big leap for Sam and calls for skills (e.g., commercialization experience) that he does not have. It also seems to call for someone who has spent their career in cell-based therapeutics, and Sam can claim to have been in that field for only seven years. Sam can make himself into a very strong candidate if he pays attention to all of the guidance that has been presented in this book. This includes 1) emphasizing his strengths and unique skills in his résumé, cover letter, and interviews, 2) downplaying the areas where he is less qualified, and 3) deeply researching the company so he can explain how he can help them.

The underlined and numbered elements in the posting are the ones that Sam needs to focus on. The following item numbers correspond to the numbered underlined items in the position description in Appendix 3:

1. Sam cannot claim to have brought a product through commercialization. On the other hand, few candidates will be able to claim this, especially if they have worked primarily in the area of cell-based therapies (fewer than forty cell-based therapies have *ever* been approved for commercialization as of 2025). Sam will not call this out explicitly and will emphasize the work he did to bring product candidates through pre-IND readiness.
2. Sam can claim a "deep understanding" of cell therapies by virtue of having been an instructor in two cell therapy courses, listed in his résumé.
3. Sam's résumé documents cross-functional collaboration and management of complex projects.
4. Sam's résumé documents his scientific expertise and leadership skills. His "business acumen" could be a weak point, but he is not interviewing for a business development job. He could address this by citing his close interactions with people in business development and marketing and emphasize his focus on strategic and commercial considerations when choosing which projects to advance.
5. Recruiting the right staff is a key role for positions at this level. Sam needs to be able to describe a rational data-driven approach to identifying and hiring staff.
6. Sam's résumé documents his ability to work with and manage outside contractors, and he should allude to this in his interviews.
7. Sam has no documented preclinical training but has worked closely with the preclinical team. As VP he will likely be managing a head of preclinical development and should be able to demonstrate a high-level understanding of the advantages and pitfalls of specific preclinical models.
8. Sam has no experience in planning or execution of clinical studies, which would normally be managed by the head of the clinical or medical team. Sam will not address this unless asked, and then he would need to talk about how his team will work with clinical to provide support and data for their work.
9. Sam needs to describe a proactive approach to keeping leadership involved and informed about R&D progress and plans.
10. Contingency planning means that if one program lags, they will have options at the ready to fill the gap. Sam can address this by outlining a systematic approach to updating and prioritizing the company's roster of R&D projects at various stages of advancement.
11. While Sam does not have a long career in cell therapies, he does have highly relevant recent experience and should capitalize on that.
12. This as well as 13 and 14 stress the need for highly developed emotional, interpersonal, and social intelligence. Sam can clearly demonstrate his recognition of the importance of these skills by having attended two intensive management and leadership training programs, as well as his ability to speak about what he learned.

A winning cover letter

Now that Sam has deciphered the position description and identified what he needs to highlight, he is ready to write his cover letter. Sam's letter is shown in Appendix 4. It addresses the most important elements of the position description and candidate requirements in a bit more than one and a half pages. A key element to note in this letter is that it has a paragraph devoted to Sam's managerial and leadership skills, including his participation in two management training programs. This training will make his application stand out. The statement at the top of page 191, *taken from Sam's letter*, is an eloquent summation of Sam's skills in this area. This cover letter should get XYZ's attention and lead to an interview, which is Sam's next hurdle.

BEYOND THE JOB DESCRIPTION: A FIELD GUIDE TO MANAGERS AND LEADERS OF SCIENTISTS

Sam should expect to be asked specific questions about his managerial and leadership skills. To prepare for these questions he needs to know and be articulate about the skills that they are looking for.

The following are ten key managerial and leadership characteristics that skilled interviewers look for. Many of these are based on principles articulated in specific chapters of *Lab Dynamics*: Chapter 5 for coaching and mentoring, Chapter 3 for negotiation, Chapter 6 for managing performance, and Chapter 4 for hiring.[3] Being able to talk about these skills during an interview is one of the clearest signals Sam can send that he takes a thoughtful approach to the roles of managing and leading. Appendix 5 contains questions Sam may get asked during an interview to assess his competence in each of these skills as well as suggestions for how he can develop his own answers to each question.

1. Coaching and mentoring

The best scientists in managerial or leadership roles are also skilled coaches and mentors. As a leader or manager, you will have people reporting to you (your direct reports). In an academic setting, these include faculty as well as other managers or administrators. In industry, many or most will be working scientists, as well as those in the roles of directors, supervisors, or team leaders. If you are a skilled coach and mentor, these people will see you as an ally and trusted adviser who has an interest in them as people and in helping them to grow and advance in their careers. Staff who see you in this way are loyal and motivated because they feel seen and valued.

2. Negotiation and conflict management

The tools covered in Chapter 9 for negotiating a job offer are also essential for much of what you do as a manager or leader. If your idea of managing is telling people what to do, it is time to learn more effective and enlightened approaches, especially when managing those who are themselves in managerial roles. Being able to resolve differences using the tools of negotiation is a skill that will enhance nearly everything you do.

You can get started learning these skills by reading *Getting Past No* by Bill Ury, which will tell you 90% of what you need to know.[60] I recommend that anyone planning to seek a senior-level management or leadership position become conversant with the language of negotiation beyond simply referencing the overused but still essential "win-win" strategy. Become familiar with the nature of underlying interests (the hidden motivations that underlie what people say they want but which they are not talking about), and during an interview be able to conjure up an anecdote illustrating how you used those interests to get to the root of a thorny problem.

3. Managing performance

Managing the performance of your direct reports is one of the most important roles of a senior manager or leader. Key skills associated with this include knowing how to set clear goals and how to give useful feedback. This is a topic that skilled interviewers will probe. Be prepared with answers that show that you understand how to set actionable goals with your reports and that you are skilled at providing honest and helpful feedback. *Let's Talk* by Therese Huston is an excellent general introduction to effective feedback.[97]

4. Managing up

From an organizational perspective, every manager or leader is sandwiched between those they report to and those who report to them. Even the CEO is responsible to the Board. Navigating the dynamics of your relationships with those above and below you is complex, but it is especially so with those above you. Managing up can be particularly challenging if you have moved from an academic role to a managerial role in an industry setting. Academics are used to a degree of freedom in how they go about their jobs that is absent in industry. Finding that a plan or decision you have made is subject to oversight or even reversal by your boss can be a jarring experience. Learning that your boss expects you to read their mind, just as you used to expect your postdocs to read yours, can be a rude wake-up call for someone used to operating in the oversight vacuum characteristic of academic institutions.

In both academia and industry, managers feel a sense of loyalty, responsibility, and possibly protectiveness to their reports and to their team. Yet at times they may be asked to make a decision that may negatively affect the team. Candidates interviewing for managerial or leadership positions need to make it clear that they understand that such things will happen and that they will need in every case to do what is best for the organization. They also need to convey that they understand the importance of knowing their boss's priorities.

5. Managing to a deadline

There are deadlines in both academia and industry. In academia the one almost everyone is familiar with is getting a grant application in on time. While that requires that *you* be a disciplined writer, it also may require getting crucial supporting experiments done by *others* for inclusion in the application as preliminary data. In industry, if you are in a public company that has announced that results from your group will be available by Q3 and you fail to deliver, your group may be responsible for a drop in share price. If you are in a startup short on cash that needs to raise money from investors, failure to deliver timely proof of concept for your platform can have similar or worse consequences.

This is another instance in which moving from academia to industry introduces pressures and constraints that take getting used to. If you are making that move it is incumbent on you to convey your understanding that time is money and that finding a solution today that is good enough to solve the problem is better than finding a solution next week that is perfect. As a managerial candidate, make it clear that you know how to help people set goals that are aggressive and realistic both in terms of objectives as well as timelines.

6. Hiring

Questions about how you go about identifying and hiring people for your team are almost obligatory when being interviewed for leadership or management roles (more on this below). If you have not already done so, read up on current best practices in hiring.[3,98] Chapter 6 of *Lab Dynamics*, written from the perspective of the candidate, is a guide to the skills you need to be looking for in new hires. Of special importance are the "behavioral" interview questions in Chapter 7 of this book. These are designed to identify candidates with desirable personal traits, including the traits listed in this chapter. Being able to rattle off a short list of traits like conflict management, emotional regulation, and feedback seeking can single you out as an outstanding candidate for a senior managerial or leadership role.

7. Managing uncertainty

One of the greatest challenges of being a manager or leader is helping your team or organization through times of uncertainty.[99] Anticipated changes in grant or corporate funding or the failure of a key corporate or grant-funded program are stressful for both leaders and employees. Your ability to help your team and organization navigate such uncertainty can be the deciding factor in your and their ability to persevere and regroup. When being interviewed, demonstrate that you understand the need for leaders to maintain a reality-based optimism in the face of uncertainty without making unrealistic or false promises.

8. Managing silos

Being able to think and talk on an organizational level are hallmarks of successful managers and leaders. This includes the ability to look beyond your group or department and to evaluate ideas and opportunities from the perspective of the organization as a whole. It means foregoing parochial concerns in favor of global ones and being able to see silos, organizational tribes that exclude others, for what they are: damaging to the overall mission.[100,101] Being able to discern and manage silos requires negotiation, diplomacy, and empathy. Silos can signal the failure of leaders to encourage collaboration, a corporate focus on department-centric success metrics, and even encouragement by managers who actively disparage other groups. Approaches that can counteract silo formation include actively promoting internal collaboration by making it an organizational value and recognizing and rewarding those who work across departments.[102]

9. Managing failure

Like uncertainty, failure and disappointment are built-in features of both scientific and administrative life. When things go wrong, as they will, organizations need managers and leaders who can inspire others and convert disappointment into action in order to remobilize. This requires having contingency plans ready for when such a time comes. Just as important is your ability to manage your own and others' disappointment, not through denial of failure but by embracing it as a learning opportunity. During an interview be ready to summon up one or two stories of how you overcame an adverse situation. Showing your awareness that this is an essential skill will signal that you are focused on the well-being of your team.

10. Managing separations

Some managers and leaders have blind spots when it comes to recognizing that a member of their team is not working out. This is understandable, since having to let go of an employee who is not performing as needed can be traumatic for the employee, difficult for the employer, and fraught with innumerable administrative hurdles. Chronic underperformers who are not let go are often shunted into roles where they can do the least harm but in which they are not doing much good either.

Knowing how to recognize when an employee is on a downward performance trajectory is a key managerial and leadership skill. Conveying to others that you are attuned to performance issues, that you provide prompt and clear feedback, and that you monitor performance for signs that the feedback has been acted upon, signal that you are attuned to this important issue.

Appendix 5 contains sample questions you may get asked about each of the preceding characteristics or practices. Guidance on how to create your own answers follows each question.

WHAT TO READ

Being able to talk about the managerial characteristics just reviewed is a skill that will take you time to develop. An excellent way to get started is to read a variety of books on management. I have listed some of my favorites. You do not need to read every book listed cover to cover. You will be able to get most of what you need by reading selected chapters or topics from each.

- *Lab Dynamics: Management and Leadership Skills for Scientists,* Carl M. Cohen and Suzanne L. Cohen[3]
- *Leading Geeks,* Paul Glen[102]
- *On Becoming a Leader,* Warren Bennis[103]
- *The Wisdom of Teams,* Jon R. Katzenbach and Douglas K. Smith[104]
- *The Practice of Adaptive Leadership,* Ronald A. Heifetz, Marty Linsky, and Alexander Grashow[105]

- *Leading Outside the Lines*, Jon R. Katzenbach and Zia Khan[106]
- *Keeping Teams on Track,* Linda Moran, Ed Musselwhite, and John H. Zenger[107]
- *Leading Teams: Setting the Stage for Great Performances,* J. Richard Hackman[108]
- *Hidden Potential,* Adam Grant[109]

CHAPTER 12 SUMMARY

Advancing your science career may require leadership and management skills

- Even if you prefer hands-on science, most organizations structure promotions so that higher pay and responsibility come with managerial and leadership duties, making these skills essential for career growth.

Job postings often understate management requirements

- Science leadership roles may list mostly technical skills, but in reality they demand significant interpersonal, organizational, and strategic capabilities, so candidates must prepare to demonstrate them.

Read between the lines when applying

- Identify the implicit managerial, collaboration, and strategic expectations in postings, highlight your strengths in these areas in résumés and cover letters, and tactfully minimize weaker areas unless directly asked.

Become familiar with core managerial competencies

- Successful science leaders consistently demonstrate skills in coaching/mentoring, negotiation/conflict management, performance management, managing up, meeting deadlines, hiring, handling uncertainty, breaking down silos, managing failure, and addressing underperformance.

Prepare for interviews with stories

- Be ready to share concise, relevant stories in interviews showing how you have applied these skills, including how you navigate organizational priorities, foster collaboration, and respond to setbacks.

Continual learning in leadership

- Building managerial expertise is an ongoing process; keep reading—every management book has something to teach you. Take a management or leadership workshop on occasion to sharpen your skills and keep you focused on the human aspects of your job as a leader of people.

APPENDIX 1

Five "managerial and/or leadership" position descriptions

Position requirements have been organized into the following categories:

- Science-related
- Managerial/leadership/interpersonal-related
- A combination of both

1. Senior director, clinical pharmacology: Pharmaceutical company

Science-related

- Experience designing clinical pharmacology and pharmacometrics components of filing packages and associated clinical studies
- Experience with regulatory agency interactions including written responses and IND packages
- Experience with Early Development Phase, as well as End of Phase 2 and NDA/MAA regulatory
- Support of drug development programs in all phases including life-cycle management
- Broad therapeutic modality experience (large molecule, small molecule, etc.)
- Experience leading cross-functional teams
- Superior communication skills
- Ability to solve complex problems while working in a matrix environment
- Academic or clinical postdoctoral training with significant publication history

Managerial/leadership/interpersonal-related

- None listed

A combination of both

- None listed

2. Vice president, research and development 1: Biotech company

Science-related

- Ph.D. in biochemistry, chemistry, biophysics, or related field
- Cell therapy experience is a must
- Proficiency in lab methods, such as PCR, electrophoresis, cytometry, NGS library preparation, mass spectrometry, and nanopore sequencing
- Experience with transferring manual assays onto engineering prototypes or commercial robotic platforms

- Experience in development of a commercial product highly desired
- Excellent problem-solving skills and attention to detail
- Experience in data analysis software
- Experience in statistical and DOE software

Managerial/leadership/interpersonal-related

- 10+ years of experience in people and project management, with increasing roles and responsibilities
- Strong analytical, emotional, and social intelligence to navigate complex team dynamics and external partnerships
- Excellent interpersonal skills, capable of building strong relationships and communicating with stakeholders at all levels
- Proven experience *building and leading development teams* focused on milestone-based scientific discoveries

A combination of both

- Strong written and verbal communication skills, with the ability to document experiments clearly and collaborate effectively with multidisciplinary teams

3. Vice president, research and development 2: Biotech company

Science-related

- Deep knowledge of cell analysis, protein-based assays, and fluorescence microscopy
- Strong grasp of the full product development life cycle, including hardware, software, and reagents
- Familiarity with regulatory requirements (FDA, ISO, etc.)

Managerial/leadership/interpersonal-related

- Exceptional strategic thinking, project management, and interpersonal skills
- Ability to articulate complex ideas to diverse technical and nontechnical audiences

A combination of both

- None listed

4. Vice president and associate provost: University

Science-related

- A demonstrated record of research and scholarship at a high level

Managerial/leadership/interpersonal-related

- The successful candidate will have a doctorate with at least 5–10 years of experience in research administration or leadership roles within a university or research-focused institution
- The successful candidate for this position will be a strategic leader, as well as a creative and innovative builder and collaborator
- Significant leadership experience in research administration
- The ability to represent the research vision, strategies, goals, and needs of the enterprise internally, nationally, and internationally
- Demonstrated ability to foster strong collaborations with deans, administrators, and faculty, as well as government, industry, community constituents, and peer institutions
- Demonstrated experience with interpreting and managing complex agreements and authoring successful extramural research initiatives
- Strong understanding of research financial management, including budgeting and compliance
- Knowledge of sponsored program operations, including research development, pre-award proposal support, and post-award grant management, and familiarity with federal agency and foundation support programs
- Experience in technology development, patents and licensing, and industry engagement with academia

A combination of both

- Significant experience in the development of cross-disciplinary research ventures
- An understanding of the needs of research faculty at all levels and how to effectively support their research, goals, and objectives

5. Vice president, discovery and preclinical science: Biotech company

Science-related

- Builds and maintains an in-depth knowledge of preclinical research, biomarker and imaging technologies, and customers adopting our products
- Expertise in process optimization

Managerial/leadership/interpersonal-related

- Strategic mindset with the ability to translate vision into actionable plans
- Exceptional leadership and team-building capabilities
- Expertise in resource planning
- Excellent communication, negotiation, and interpersonal skills

- Proficiency in leveraging data analytics and technology to inform decision-making
- Excellent interpersonal, verbal, and written communication skills
- Exceptional teamwork and time/project management skills
- Strategic thinking and vision
- Leadership and influence
- Collaboration and relationship-building
- Change management
- Decision-making and problem-solving
- Results orientation
- Resilience and adaptability
- Customer and stakeholder focus
- Innovation and continuous improvement

A combination of both

- None listed

APPENDIX 2

Management and leadership candidate résumé

Samuel Haberman, Ph.D.
27 Grandview Ave.
Belmont, MA 02349

Haberman238@email.com | (617) 555-0198
linkedin.com/in/haberman238–biotech

PROFESSIONAL SUMMARY

Experienced biotechnology R&D leader with 12+ years in therapeutic development, specializing in cell therapy platforms. Successfully led multidisciplinary product-focused teams and created a project management structure to advance programs to preclinical readiness. Proven track record assembling and managing teams of 60+ employees across multiple therapeutic areas. Strong background in protein engineering and biologics development with recognized expertise in cellular therapeutics.

CORE COMPETENCIES

Management Experience: Team Leadership • Project Management • Budget Oversight • CRO Management Experience • Cross-functional Collaboration • Strategic Planning
Scientific Expertise: Cell Therapy—multiple platforms • Protein Engineering • Biologics Development • Cell Culture • Molecular Biology • Biochemical Assays • Bioconjugation • Immunology

PROFESSIONAL EXPERIENCE

NORTHEAST BIOLOGICS | Burlington, MA

Director, Cell Therapeutics | *2019–Present*

Planned and created company's first cell therapy program and currently lead development activities for entire cell-based therapeutics portfolio. Overall responsibility for management of 60-person team of 15 scientists with $16M annual budget. Advanced 3 biologics programs through preclinical development, with 2 programs reaching IND-enabling study stage.

- **Program Management:** Designed and implemented company's project management system for projects in cell therapy including CAR-T and osteogenic therapeutics.
- **Team Leadership:** Built cell therapy team from inception. Maintained team morale and stability during 1-year company retrenchment.
- **Regulatory Support:** Responsible for generating supporting data package for pre-IND-FDA meetings, presenting to FDA, and crafting regulatory strategy with VP of Regulatory Affairs.
- **Process Development:** Established in-house capacity for manufacture of cellular therapeutics, working with CROs for specialized phenotypic and genotype testing.

(Continued)

CAMBRIDGE THERAPEUTICS | Cambridge, MA

Senior Scientist, Discovery Biology | *2015–2019*

Led target validation and assay development efforts for oncology-focused small molecule and biologics programs.

- **Assay Development:** Led 3-person team that developed 8 novel cell-based assays successfully used by screening team to identify over 30 lead compounds in a 3-year time span.
- **Target Validation and MOA Biology:** Lead scientist responsible for identifying the mechanism of action studies for 2 small molecules allowing them to move into lead optimization.
- **Management:** Supervised 3 junior scientists while contributing to cross-functional project teams.
- **External Partnerships:** Lead company liaison responsible for relationships with 2 academic labs focused on advanced fluorescent imaging for assay read-out.

BIOTECH SOLUTIONS INC. | Woburn, MA

Scientist II, Cell Biology | *2012–2015*

Advanced company programs by developing cell-based assays supporting 3 preclinical programs.

- **Assay Development:** Lead scientist in team that developed 3 novel in vitro assays for therapeutic compound evaluation.
- **Data Analysis:** Introduced automated data analysis tools to visualize key attributes of candidate compounds for project team review.
- **Methods Development:** Adapted legacy protocols for use in new cell lines and assay conditions.
- **Collaboration:** Co-lead in cross-functional working group with medicinal chemistry and pharmacology teams.

UNIVERSITY OF MASSACHUSETTS | Amherst, MA

Postdoctoral Research Associate | *2009–2012*

Part of a 3-person project team under Dr. Jennifer Walsh that identified a novel signaling pathway critical for the function of the AKT tumor suppressor gene.

- **Research Focus:** Identified molecular mechanisms of tumor suppressor proteins using biochemical and cell biological approaches.
- **Publication:** Published 3 first-author papers in peer-reviewed journals.
- **Collaboration:** Worked with structural biology core facility for protein characterization studies.

(Continued)

EDUCATION AND TRAINING

Ph.D., Biology
University of Connecticut, Storrs, CT | *2009*
Dissertation: "Role of Protein Modifications in Cell Cycle Regulation"
Adviser: Dr. Robert Kim
B.S., Biochemistry
University of Massachusetts, Amherst, MA | *2004*
Management Training

- "Leadership in Bioscience": Four-day competitive-entry workshop at Cold Spring Harbor Laboratory, focused on management, leadership, and interpersonal skills development for emerging science leaders | February 2019
- "Leading High Performing R&D Teams in Biotechnology": Massachusetts Biotechnology Council, 3-day intensive workshop | June 2020
- Participant: "Developing a Regulatory Strategy for Biologics": FDA Workshop, 2021

PUBLICATIONS AND INTELLECTUAL PROPERTY

Publications: 32 publications in peer reviewed journals, 28 conference abstracts, and 1 book chapter in fields including cell-based therapy and mechanisms of tumor suppression. Full bibliography upon request.
Conference Presentations: 8 poster presentations at AACR and regional biotech meetings
Patents: 2 issued patents in protein engineering, 1 pending application in assay development

PROFESSIONAL AFFILIATIONS

Professional Organizations:

- American Association for Cancer Research (AACR)—Member
- American Chemical Society (ACS)—Member
- American Society of Gene & Cell Therapy (ASGCT)—Associate Member

PROFESSIONAL HONORS AND RECOGNITION

- Instructor: "Advances in Cell Therapy Manufacturing"—BIO Industry Workshop, 2023
- Instructor. "CAR-T Cell Development Fundamentals"—Online EMBO Course, 2022

ADDITIONAL INFORMATION

Languages: English (native), German (fluent)
Continuing Education: Currently enrolled in online certificate program in "Cell & Gene Therapy Development" through UC San Diego Extension

APPENDIX 3

Management/leadership job description

XYZ Therapeutics Vice President, Research and Development	
Position description *Underlined and numbered text refers to items that need to be addressed in the cover letter or interview.*	***What it means***
XYZ Therapeutics is a preclinical biotechnology company based in Cambridge, MA, applying proprietary insights to develop and commercialize cell therapy products addressing significant opportunities with unmet clinical needs. Building on work from leading labs, XYZ has strong IP positioning and is backed by early investors and founders of successful high-profile biotechnology companies.	
Our innovation-driven company is looking for an experienced and visionary Vice President, Research & Development to lead our team and advance our technology. This strategic leader will be responsible for (1) overseeing the development and execution of novel strategies, driving product development from concept through commercialization, and ensuring alignment with our company's growth objectives.	*The inclusion of "visionary" could give you pause since it is hard to objectify. Nonetheless, finding something in your work experience that qualifies as thinking outside the box or being an early adopter of a novel idea or technology is a good start. Another high bar is having driven a project through commercialization, which very few candidates can claim to have done.*
(2) The ideal candidate will bring a deep understanding of cellular therapeutic systems, with a proven track record of successful R&D leadership in the biotechnology sector.	*These deserve attention, especially the focus on cellular therapeutics, which is likely a must-have item. The "proven track record" part is very flexible. There are many ways you could frame your background to comply with this, and you should do so. Also note that this refers to the "ideal" candidate, who may not exist. You may be able to get by with some or most of this.*
They should possess exceptional strategic planning skills, a strong ability to (3) foster cross-functional collaboration, a demonstrated aptitude for managing complex projects, and a history of driving innovation.	*Almost any example of having contributed significantly to the creation of a new company program can be framed as involving strategic planning. Also, as above, you should be able to find examples to demonstrate a "history of driving innovation."*

(Continued)

This role requires a (4) blend of scientific expertise, business acumen, and leadership skills to navigate the evolving therapeutic landscape and deliver groundbreaking solutions that enhance patient outcomes.	*The "blend of skills" is very important and should be addressed in the application letter.*
Key responsibilities: Arranged into categories	***What it means***
Management and oversight • Oversee and manage the R&D pipeline, including discovery, research, preclinical, and clinical development, ensuring alignment with corporate objectives.	*Nothing surprising here. The inclusion of clinical development could be considered outside the boundaries of a VP R&D, belonging instead to a VP of Clinical Research or Chief Medical Officer. "Oversee" is the key phrase here, since the actual work will be done by your direct reports.*
• Provide oversight for establishment of research scale processes and associated analytical methods.	
• Oversee the R&D budget, allocating resources effectively to prioritize key projects and ensure efficient use of funding.	*Be ready to acknowledge that this requires close collaboration with the Chief Financial Officer (CFO).*
• (5) Recruit, lead, and mentor the R&D team, comprising scientists, engineers, and technicians.	*This is one of the three or four most important elements in this list. Being able to explain how you recruit and mentor will be essential.*
• Actively work to promote team building and morale within the department.	*You will need to discuss how you foster and promote a team mindset in your staff. Requires an understanding of the need for both team and individual recognition and rewards. Have examples of your ability to manage external contractors.*
• Effectively manage all aspects of the research and development effort including participation in program teams.	
External relationships • (6) Establish and maintain effective relationships with CROs and consultants for the design and execution of nonclinical studies.	
• Collaborate with internal and external stakeholders, including the Regulatory, Manufacturing/CMC, Quality, and external consultant to determine new product development strategy from product inception to market launch.	*If you can provide examples of working with other company functions, especially manufacturing and marketing, this will be a plus. Find out how the functions listed here fall within the company's org chart and which if any report to you.*

(Continued)

• Engage with clinical Key Opinion Leaders, describing the platform, preclinical, and clinical opportunities to enable patient recruitment.	*Patient recruitment implies that they expect to be in the clinic soon.*
• Maintain a professional and credible image with key physicians, consultants, suppliers, and coworkers.	
• Develop research proposals to collaborate with institutions, cooperative groups, and pharmaceutical companies on development studies.	
• Represent the organization at scientific conferences and events.	
Hands-on involvement • (7) Employ preclinical models to establish exposure, biomarker, and efficacy relationships.	*The word "employ" should be read as "Lead the development and use of . . ."*
• (8) Build and execute translational plans that allow patient selection and stratification, early markers of clinical activity, and provide key mechanistic insights.	*Except for the "mechanistic insights" part, these functions would most likely lie with others, including the head of clinical research.*
Big picture • (9) Maintain communication with senior management team regarding program progress, risks, challenges, and opportunities.	
• (10) Develop contingency plans and backup scenarios to proactively mitigate unforeseen circumstances.	*In this context, contingency planning means that you have a list of pre-vetted R&D opportunities that you can call upon if one program or approach fails to show promise.*
Candidate qualifications	***What it means***
Scientific skills • Ph.D. in biochemistry, chemistry, biophysics, or related field with at least 15 years of experience is preferred.	*The key word here is "preferred." This means that these qualifications are very flexible as indicated by the "related field." It is not even clear whether a Ph.D. is required although it would be unusual if it were not.*
• (11) Cell therapy experience is a must.	*While cell therapy experience is a "must have" they do not specify what this means. You could frame almost any amount of work in this area as "experience."*

(*Continued*)

• Experience in development of a commercial product highly desired.	*The word "desired" is a tip off that this is a "nice to have" since there are no qualifiers like "documented" or "extensive."*
Managerial and people skills • (12) 10+ years' experience in people and project management, with increasing roles and responsibilities.	*This is important, but there is a lot of flexibility in the term "experience."*
• (13) Strong analytical, emotional, and social intelligence to navigate complex team dynamics and external partnerships.	*The inclusion of "analytical" with emotional and social intelligence is a bit confusing (suggesting an AI author or lax editing). However, their calling out social and emotional intelligence is significant.*
• (14) Excellent interpersonal skills, capable of building strong relationships and communicating with stakeholders at all levels.	*For a leadership role, these are arguably among the most important skills in this list. You should be able to frame much of what you have done as people and project management and be able to give specific examples.*
• Proven experience building and leading development teams focused on milestone-based scientific discoveries.	*Another reference to team building, signaling their focus on the importance of the interpersonal skills of the candidate.*
Other • Willingness to operate with urgency, be resourceful, and maintain flexibility in an evolving environment. • Ability to work independently and collaboratively in a highly dynamic work environment. • Strong written and verbal communication skills, with the ability to document experiments clearly and collaborate effectively with multidisciplinary teams.	

APPENDIX 4

Cover letter for management/leadership candidate

Samuel Haberman, Ph.D.
Haberman238@email.com | (617) 555-0198
Belmont, MA 02349

Date

XYZ Therapeutics Hiring Committee
XYZ Therapeutics
97 Claflin St.
Cambridge, MA

Dear Hiring Manager,

I am writing to express my interest in the Vice President, Research & Development position at XYZ Therapeutics, advertised in *Science Careers*.

I am a Ph.D. cellular biologist with over twelve years of R&D leadership experience in the biotechnology industry, most recently focused on cell-based therapies. I am eager to contribute at both the scientific and organizational level to further XYZ's exciting programs and believe that my combination of scientific expertise and managerial skills will enable me to help advance your lead programs toward the clinic.

My scientific expertise encompasses protein engineering, biologics development, and cell-based therapeutic platforms. At Northeast Biologics, I was responsible for leading our entry into cell therapeutics, including the creation and staffing of our multidisciplinary CAR-T therapeutics team and establishing productive partnerships with two leading academic laboratories. These efforts have enabled us to advance our lead program in osteogenesis into early preclinical testing. This experience has provided me with extensive hands-on expertise in cellular therapeutics, regulatory submissions, and a deep familiarity with the unique challenges of the cell therapy space. I was selected to be an instructor at two professional training programs including the BIO "Advances Cell Therapy Manufacturing" workshop as well as the EMBO workshop "CAR-T Cell Development Fundamentals." My ongoing enrollment in the Cell & Gene Therapy Development certificate program through UC San Diego Extension demonstrates my commitment to keeping current with advances in this critical field.

I believe that a key to success in positions such as this one is an ability to see beyond the science to the people involved and the dynamics of scientific teams. I am skilled at envisioning

(*Continued*)

opportunities where others see problems, building alliances where none seem possible, and resolving conflicts when no agreement seems evident. I am able to bring organization and focus to scientific endeavors that by their nature are unpredictable. I have expanded my management and leadership skills by attending training programs specifically for managers and leaders in science, and I plan to continue to do so. I am equally committed to mentoring my direct reports to help them develop into effective managers and leaders. My focus on management has enabled me to create inspired R&D teams whose members feel valued and appreciated. I am keenly aware of the complexity of leadership roles, especially in the need to balance the interests of internal and external stakeholders.

These skills have enabled me to build and manage cohesive, high-performing scientific teams. At Northeast, I built the Cell Therapeutics Group from the ground up and currently lead cross-functional teams of 60+ employees with a $16M annual budget. I implemented our current project management system for tracking multiple preclinical programs, and I currently manage five CRO relationships and multiple academic partnerships. I have been actively involved in crafting our regulatory strategy and have participated and presented at two pre-IND meetings with the FDA. I am gratified at having successfully advanced three biologics programs through preclinical development, with one program reaching IND-enabling study stage.

I am excited about XYZ's innovative approach to cell therapeutics and your strong IP position in the cell therapy space (especially your platform-enabling "Ingersol et al 2021" patent for patient specific therapeutics). Your focus on applying proprietary insights to address unmet clinical needs is an industry paradigm, and I am eager to become part of XYZ's efforts to advance therapies that meaningfully impact patient outcomes. XYZ's leadership team, strong financial backing, and relationships with leading academic labs represent the ideal environment for driving breakthrough innovations in cellular therapeutics.

I am confident that my track record of building successful teams, advancing complex programs through development milestones, and fostering innovation within resource-constrained environments will enable me to advance your R&D objectives.

I would welcome the opportunity to discuss how my experience and passion for cell therapy can contribute to XYZ Therapeutics' continued success. I look forward to hearing from you and to the possibility of joining your team in advancing these promising therapeutic approaches.

Sincerely,

Samuel Haberman

Samuel Haberman, Ph.D.

APPENDIX 5

Sample management/leadership interview questions

1.	**Coaching and mentoring** "Have you ever had an employee who was not performing up to your requirements? How did you handle that? How did it turn out?" "Have you ever had an employee who was a very high performer? How did you acknowledge that?" Talk about your ability to provide useful performance feedback that helps an employee improve and your awareness that employees need and want feedback. Also important is recognition that positive feedback can be as or more helpful than feedback that focuses only on needed improvements. Point out that frequent and early feedback can and often does prevent performance issues from becoming problems.
2.	**Negotiation and conflict management** "In your previous roles have you ever had a situation where you had a disagreement with one of your reports or with your boss? Tell me about that and how it turned out." "Have you had a time when two or more people reporting to you had a disagreement? How did you handle that?" Show that you recognize that such conflicts or disagreements are natural and can be dealt with productively. Have a story that illustrates how you helped mediate a solution in a conflict between others. Be aware of a tendency to speak in terms of whose "fault" it was.
3.	**Managing performance** "How do you go about letting people know what is expected of them?" "Have you had a case where someone was not performing up to your expectations? How did you manage that?" Tell a story that shows you know how to set goals and how to manage expectations. Talk about how you do that—do you have frequent team and one-on-one meetings where you provide feedback and guidance? If you have a story about a low performer who you helped by giving them feedback or advice before their performance became a problem, now is the time to tell it.
4.	**Managing up** "Have you had occasion to disagree with a decision or request made by someone senior to you? What did you do?" Show that you have a willingness to engage authority in a collegial manner about differing views and that you have the ability to "choose your battles." Since nearly all organizations are hierarchical to some degree, show that you understand that sometimes people in authority can and should call the shots.

(Continued)

5.	**Managing to a deadline** "Have you had a situation where you needed to get something done by a deadline and were concerned that those responsible for the work could not or would not be able to make that happen? What did you do?" "Have you had to manage someone who failed to deliver on time on multiple occasions? What did you do?" Show that you can clearly convey priorities to employees and sense which ones need help in developing strategies to meet deadlines and prioritize work. Have a story about an underperforming employee where you coached them to become more responsive.
6.	**Hiring** "Have you hired people? Tell me how you go about that and what you look for." "Have you ever made what turned out to be a bad hiring decision? Tell me about that." Make sure you understand the principles of hiring using a methodology as opposed to "gut feeling" (see *Lab Dynamics*, Chapter 4). Emphasize that you know the importance of finding staff with a good balance of technical skills and valued personal traits. Describe how you make sure you have input from other team members in hiring decisions.
7.	**Managing uncertainty** "Have you been in a situation where you were unable to give your team the information or guidance they wanted or needed because of uncertainty within the organization? How did you manage that?" "Have you ever felt that as a manager you lacked crucial information about a project or program in your organization? How did you manage that?" Either have a story or describe a situation in which you were able to keep your team motivated despite uncertainty. Demonstrate your ability to manage team members' resentment or "acting out" against senior management in times of uncertainty.
8.	**Managing silos** "In your previous positions, were there silos or organizational domains separating different parts of the organization? Did this affect your ability to get work done? How did you manage that?" Show your awareness of how silos can hamper information sharing and collaboration. Talk about how you work to build bridges and form alliances with other parts of the organization and how you encourage others to do the same.
9.	**Managing failure** "Have you had a project or program fail? How did your team or group take it? What, if anything, did you do or say to help them?" Demonstrate that you have a willingness to confront "failure" openly and to learn from it and that you appreciate that employees need closure and support when projects fail or need to be stopped for a strategic reason.
10.	**Managing separations** "Have you ever had to let someone go? Tell me about that. How did it come about?" Have a story about your ability to pick up on and deal with poor performance early before it becomes a problem. Mention your awareness of the important role of getting an HR specialist involved early rather than waiting for performance failure. If you have had to let someone go, talk about what you learned from the experience.

13 Meet the CEO

- **Questions from the CEO**
  - Tell me about yourself
  - Self-knowledge, weaknesses, motivations: Listening for more "we" than "I"
  - Managing adversity
  - Why are you applying for this job?
  - Your management and leadership style
  - Mind games
  - Your questions and ideas for them
- **CEO interview insights: What really matters**
  - Core qualities CEOs value
  - Practical advice
- **Other factors that can work in your favor**
- **Contract negotiations for management-level jobs**
- **Chapter 13 summary**

The previous chapter began your preparation for management/leadership position interviews. Many of the people interviewing you will be nonscientists such as human resources specialists and people in business development, finance, and marketing. They will not be evaluating your scientific expertise; they will want to get to know you as a person and assess your managerial and leadership skills. These will include the characteristics listed in the previous chapter as well as communication skills, empathy, and self-awareness. The best and most experienced interviewers, like professional recruiters, are pretty good at identifying people who have such skills.

The most important person you will be meeting in this group is the CEO. In nearly all organizations the opinion of the CEO will be the deciding factor in whether you get hired, sometimes regardless of what anyone else thinks. CEOs use a wide array of questions, rubrics, and general rules to judge who is going to work out, at various times relying on instinct, experience, and "gut feeling." The question for you is whether there are common themes that you can prepare yourself to address when you meet the CEO. The answer is yes; to show you how, I am going to introduce you to the former *New York Times* writer Adam Bryant.

QUESTIONS FROM THE CEO

Every Sunday between 2009 and 2017 Bryant wrote a column in the *New York Times* called "The Corner Office."[110] Each featured an interview with a CEO. The kinds of companies represented ran the gamut from pharmaceutical companies to house cleaning, restaurant chains, web services, and banks. Each week, Bryant's questions for the CEOs remained pretty much the same, so by reading these interviews you could get a picture of how different CEOs thought about the same things. The questions included "Tell me one lesson you have learned about leadership" and "What is your leadership style?" For our purpose, the most interesting questions were "How do you hire?" and "What questions do you ask during job interviews?" As I read the CEOs' answers to these questions each week, it seemed that everyone was asking candidates different questions and that every CEO also had their own "make it or break it" question to ask. It seemed like there was no commonality in how these CEOs interviewed—there was no playbook.

I decided to test this hypothesis by rereading nearly all (there are over four hundred) of the interviews in the *New York Times* archives. I looked for differences and overlaps in the questions the CEOs asked when interviewing a candidate. What I found surprised me. While each CEO seemed to think that they had their own "secret sauce" for identifying winners, the reality was quite different. In sharp contrast to the impression I got from reading one interview each week—that there was no overlap—when I read them one after another, I found a lot of overlap. In the following, I have collected the most common interview questions CEOs said they asked and have organized them into categories.

The parenthetical comments in quotes following some of the questions are quotes from Bryant's CEOs about what they are looking for by asking that question.

Tell me about yourself

The CEOs said they wanted a sense of "who you are." As noted earlier, you need to have a succinct one- to three-minute story. You need to tell it like you are putting it together at that moment—i.e., it should not flood out like a speech and should not sound rehearsed. Like your résumé, your story can and should be customized for each person, company, or interview. Decide what aspect of your experience needs to be emphasized based on the position you are applying for and the focus and needs of the organization. If you do not know this in advance, listen hard for clues during the interview and especially when you get a chance to ask them questions. Typical questions include the following:

- You've got three minutes. Tell me about yourself.
- What were your goals when you were growing up?
- Walk me through your résumé, particularly why you changed from one job to the next.
- Tell me three things about how you define yourself.
- What magazines or papers do you read in your spare time?

Self-knowledge, weaknesses, motivations: Listening for more "we" than "I"

CEOs asked the questions below to get a sense of how self-aware you are. More than 50% of CEOs ask some version of "Tell me what you don't do well" or "What's your biggest weakness or deficit?" (discussed in Chapter 7). They say they are looking for people with a "growth mentality."

- When things aren't going well, how do you get yourself motivated?
- What are you good at and why do you think so?
- What irritates you?
- What matters most to you?
- If you failed at this job in your first ninety days, what things wouldn't you be doing well? What don't you know, but know you need to know, in order to be successful at this job?
- What are you genuinely bad at? What does your spouse or partner or the person you are dating say you are bad at? (*"Because if they haven't told you, then you shouldn't be sitting here because I can't work with you if you don't know what you're bad at."*)
- Tell me about the first experience in your life when you realized that you had the power of change or the power to do something meaningful.
- What are you working on developing within yourself right now? (*"If someone can't answer that honestly and rather quickly, if you have to think too hard about what you are working on developing, are you really working on it? I'm looking for a commitment to self-development and personal growth."*)
- Take me through what your past employers said to you in your reviews about what you do well and where you need to improve.
- What gets you out of bed? What gets you excited? What are you passionate about?
- Tell me about a time on a Sunday that you were thinking about going back to work and you hated your job. Why? What was it? (*"That tells me a lot about the type of team member you are going to be."*)
- What kind of cultures do you like to work in?
- What do you think would be challenging for you to be working in this environment?
- What will your company/organization think when you tell them you are leaving? (*"It can help you get at other aspects of the person—how they work with others and how they are perceived."*)

Managing adversity

The CEOs said they were looking for evidence that you can admit failure and learn from it. For most CEOs, how you respond to adversity or failure is a key indicator of strength of character and perseverance. Have a failure or adversity story that you can tell. The take-home message should be "I'm not perfect, but I learn from my mistakes." Make sure you take ownership for your own failures—blaming others will invariably be viewed as a sign that you cannot take responsibility for your own actions.

- Tell me about a time you experienced a really difficult situation. I don't care whether it happened at work or outside of work. What did you do?
- What do you do if you don't agree with a decision?

- Tell me something you did that was really hard and that you didn't handle as well as you wish you had. (*"Part of being a leader involves a willingness to step up and do things that are hard. You don't always do that perfectly. It's precisely because they are hard that they require some willingness to learn and get better."*)
- Tell me about a time when you took a risk and failed. (*"I have not hired people who have told me they've never failed. You don't learn if you don't fail."*)

Why are you applying for this job?

The CEOs want you to show them that there is a strategic reason why you are applying for this position. They want applicants to know their company's mission, to love what the company does, and to show a drive to contribute to it.

- Why are you interested in the role? How has your career path brought you to this point? Where have you been, and where are you going?
- If this interview were reversed and I was interviewing for this job, what would you be looking for from me? (*"That helps me understand their aspirations for the job."*)
- Tell me what you think this job is all about. (*"I think just allowing them to speak about what they want to do, and what they think the job is about, is actually very useful, because it sheds light on what they think they want to do in the company. I typically also end an interview with asking them how they can make a difference."*)
- Why are you leaving your current job?
- What concerns you about our organization? What makes you feel a little bit afraid? What is it that you are not sure about—whether you can meet certain expectations or maybe something you're not sure you'd like to be a part of?
- What are the qualities of a job you are looking for? (*"Do they talk about the workplace? Do they talk about the kinds of people they want to work with? Do they talk about what kinds of responsibilities they want to have?"*)

Your management and leadership style

Experienced interviewers, and CEOs in particular, want you to be able to talk about your management and leadership style. If you do not have that language at the tip of your tongue, read the previous chapter and some of the books listed at the end of it.

As we noted earlier, some CEOs said that they listen for how much you say "I" did this or that instead of "we" as an indicator of whether you see yourself as an individual player or a team player. If you are applying for a position as a bench scientist by all means use "I" to highlight all the skills that you have and the depth of your knowledge. If you are applying for a management or leadership position, using "we" will be seen by some as a tip-off that your thinking is team focused.

- How do you make hiring decisions?
- How do you lead and manage—give me an example of each. (*"I want to understand how you think about managing people who are struggling and how you bring out their best."*)

- Pick a project and tell me about it. (*Listening for use of "I" vs. "we."*)
- Talk to me about one of your greatest achievements. (*"I'll listen to the story from the perspective of whether it was an "I" and "me" achievement or a "we" achievement."*)
- Tell me who your favorite boss was, and why, and who your least favorite boss was and why. (*"You quickly get a sense of what leadership styles work best for them."*)
- Tell me about a time when you really had to stick your neck out for the greater good of the mission. (*"You want to see whether people were willing to take bold action to move the mission forward."*)

Mind games

Some CEOs fancy themselves amateur psychologists and ask questions that they believe reveal the inner workings of your mind or your motivations. Read through the questions and practice answering them: forewarned is forearmed.

- What do you want to talk about? (*"... always my first question. I get to understand how you think with that one question. I will understand your priorities and the natural flow of your thought process. I'm looking for someone who is naturally curious and has a good business mind and is asking questions to understand how to win."*)
- What would your enemies say about you? (*"If they say they don't have any enemies, I'll say "Let's pretend you do. What would they say?"*)
- Tell me how you spent your weekend.
- What did you learn from your mom?
- What percentage of your life do you control? (*"This is always a winner. It helps me understand how they feel about the world out there."*)
- If you could do anything and money wasn't an object, what would that be? What would your dream job be? Why aren't you doing that?
- What's the biggest opportunity you've missed in your professional life?
- When you are at work, what do you love to do every day? What do you hate to do?
- What makes you want to howl at the moon? What makes you really mad, so mad that you feel your pulse in your throat and you want to puke?
- What's the one thing you are doing in your job now that you hope to never have to do again?
- Tell me what the inscription is going to say on the gold watch you receive from your last company when you retire. (*"It allows me to understand truly what they really think about themselves."*)
- "I take people out to dinner and watch how they behave. Do they rush into the room ahead of others? How do they treat and respond to the waitstaff? You can tell if someone is a good person by how they treat service people." One CEO said that he would invite a candidate to breakfast, get there early, and instruct the restaurant staff to make mistakes with the candidate's order so he could see how the candidate managed adversity and treated the waitstaff.

Some of these questions may sound odd or leave you at a loss for an answer. Chances are, even if they explained what they are looking for, you would still be unsure how this question tells them that. As advised earlier, if a question stumps you, instead of answering its literal meaning, answer what you think they are trying to get at.

Your questions and ideas for them

CEOs value your curiosity about their company and the research you have done on them. Have more questions ready than you can reasonably expect to cover. Show that you have read their publications, press releases, and financials (if you understand them) and are familiar with their products. Have questions about these as well as future direction, strategy, and plans.

- You have talked to a lot of people, but is there anything about the mission of the company, our board, or my background that you want to ask me about? (*"Believe it or not, seven out of ten people say 'I've already gotten what I need.' Maybe they are nervous, but I'm thinking 'You're talking to the CEO. You don't want to know what makes me tick?'"*)
- What's the biggest, boldest, or greatest idea you have for us that we're not doing that you think would help transform our business?
- What do you think we're not doing well enough that you think you could help us do better?

CEO INTERVIEW INSIGHTS: WHAT REALLY MATTERS

For the best organizations and most thoughtful interviewers, your technical qualifications are merely a prerequisite that gets you in the door. CEOs are evaluating who you are as a person, whether they feel comfortable entrusting their scientific programs and staff to your leadership, and whether you are someone they would want to work with closely.

Core qualities CEOs value

- **Authenticity over perfection:** CEOs consistently value honesty, openness, and humility. They value candidates who demonstrate self-knowledge and emotional intelligence—the ability to understand what motivates them and gets them excited about their work each morning.
- **Growth mindset:** It is not only acceptable but often mandatory to acknowledge your imperfections, provided you understand your weaknesses and are actively working to improve. The CEOs said they look for resilience and evidence of learning from failure. If there are things that annoy you, they want you to be aware of them and to manage your reactions so that they do not impact your performance or relationships.
- **Spontaneity:** Many CEOs actively resist what come across as overly rehearsed responses. While preparation is important, over-preparation can be counterproductive, leading candidates to focus on rote answers rather than ones that come from personal experience.

Practical advice

CEO evaluations seemed to rely as much on "gut feeling" as on logic. They matched their subjective assessment of responses against their mental model of an ideal candidate. Since you cannot know this mental model, your best strategy is to build your answers on your core values and experiences rather than trying to guess what they want to hear. The most effective preparation for your interview with the CEO is to be a great person with strong scientific credentials, deep self-awareness, emotional intelligence, and an inspiring management style.

While working on these fundamentals, consider that the CEOs expect you to do these things:

- Research their company thoroughly
- Prepare thoughtful questions that demonstrate genuine interest
- Articulate specific reasons why you want to work for their organization
- Clearly explain what value you bring to them

As you develop self-knowledge through practice and reflection, you will be able to present your authentic self, highlighting the valued qualities these CEOs are looking for.

OTHER FACTORS THAT CAN WORK IN YOUR FAVOR

- **The ability to speak about what it means to manage and lead.** Just as academics can be expected to get asked "What is your approach to teaching or your teaching philosophy?" you should expect to be asked "What is your management style? How do you manage people?"

  While no one expects you to deliver a treatise on the latest in leadership thinking from Harvard Business School, they do expect that you will be able to tell them how you approach this key requirement of your new job. The simplest strategy is to have one or two well-rehearsed anecdotes to illustrate your management style. Beyond that, read some of the books I listed earlier and be able to mention one or two concepts that resonate with you from a few of them.

- **You have attended management or leadership training courses or workshops.** This is something that our management candidate Sam Haberman from the previous chapter did. If you have been lucky enough to be at an institution that offers such workshops for its faculty or postdocs and have taken advantage of these, definitely put that in your résumé. That will automatically put you in the top 10% of applicants and show that you understand the importance of acquiring these skills.

- **You have read some of the books about management and leadership listed earlier and can mention a few of your favorites.** You do not need to read all of these books in their entirety. Most management books have only a few big messages, which you can glean from reading the dust jacket or reviews on Amazon.

CONTRACT NEGOTIATIONS FOR MANAGEMENT-LEVEL JOBS

Everything we said in Chapter 11 about negotiating a job offer in industry applies here. For management or leadership jobs, there may be few or extremely broad guidelines for items such as base salary, annual bonus, signing bonus, and stock options. There will, however, be precedents with regard to others in similar roles.

If you are lucky enough to get an offer from a publicly traded company, you can find a wealth of information in the filings with the US Securities and Exchange Commission (SEC). All public companies must file comprehensive annual reports (called a 10K report) and more simplified quarterly reports (a 10Q) with the SEC. These are available to the public on the SEC's searchable online database.[111] These filings contain salary, bonus, stock option information, and much more for key company executives (the CEO, the CFO, and the three other most highly compensated executives).

If you are applying for one of the top four roles in such a company, that list can provide a benchmark for the kind of salary you could expect, adjusting for the fact that these numbers may be a year or more out of date. You can reference such numbers during a negotiation.

Beyond salary, a substantial signing bonus as well as generous stock option awards are to be expected and common. As with salary, in cases where a public company has made a new senior hire, they may need to put the terms of the new hire's contract into one of their filings, most often at the end or in an addendum or appendix. These are worth searching for since they will include valuable information you can use as guidance when you are negotiating these terms.

Negotiable noncompensation elements for senior managers and leaders

Item	Rationale
Vacation time	"I push myself incredibly hard when I'm working and I think you will see that in my productivity and results. I need periodic down time to maintain this pace, and being with my family on vacation is one way I do that." "Obviously I maintain connectivity with what's going on when I'm away."
Administrative assistance	"I maintain a very full calendar of daily meetings and expect to be traveling to visit our San Diego site regularly. Having someone who can help me manage my schedule, set up complex meetings or events, and run interference for me when my workload gets intense will be key to my being effective here." "A high-level admin will allow me to ensure that important things don't get neglected."
Office space	"I will need an office both at our headquarters as well as a smaller space on the R&D campus. I expect to be spending significant time in both places and need a dedicated space to meet with both individuals and small groups in both places. Regularly having face-to-face time with both the executive and research teams is how I keep on top of things."

For senior managers and leaders, vacation time, availability of administrative help, including a full-time or shared administrative assistant, specifically identified office space, and location and some percentage of remote work are also negotiable. The table above shows some of these noncompensation items along with ideas on how to justify what you need.

As in our discussion of negotiating academic contracts, an underlying interest of all of your requests is that you both want to create conditions in which you will be maximally motivated and equipped to succeed, and those conditions will converge to make you as productive as possible.

CHAPTER 13 SUMMARY

CEO opinion is decisive

- The CEO's assessment will typically be the determining factor in whether you get hired for a management role, often regardless of what other interviewers think, so this meeting is crucial to your success.

CEO questions follow predictable patterns

- Despite each CEO thinking they have unique interview approaches, analysis of over four hundred CEO interviews reveals common question categories: tell me about yourself, self-knowledge/weaknesses, managing adversity, why you want the job, management style, and mind games.

Authenticity trumps perfection

- CEOs consistently value honesty, self-awareness, and growth mindset over polished but rehearsed responses. They want to see that you acknowledge weaknesses and actively work to improve them.

Prepare failure and adversity stories

- Be ready to discuss times you have failed or faced challenges, taking ownership of mistakes and demonstrating what you learned. CEOs see how you handle adversity as a key indicator of character and leadership potential.

Demonstrate management knowledge

- Stand out by showing you have invested in developing management skills through formal training programs and relevant reading, and can articulate your leadership philosophy and style with specific examples.

Research thoroughly and ask strategic questions

- CEOs expect you to have deep knowledge of their company and thoughtful questions that show genuine interest. Prepare more questions than you can reasonably cover, focusing on strategy, future direction, and how you can contribute.

Contract negotiation expands beyond salary

- For senior roles, negotiate vacation time, administrative support, office space, and other elements based on how each will help contribute to your effectiveness. Use publicly available SEC filings for benchmarking compensation at public companies.

14 When to Walk Away
Red Flags and Warning Signs

THIS JOB IS NOT FOR ME

If you have applied for a job and started the interview process, you have some belief that this job is for you. Initially this was based on the job description. But as in Bayesian analysis where you update your expectations as you obtain new information, progressing through the application process yields more information about the job, the organization, and the people. So, your views may change. They may change so much that they need to convince *you* to join *them* instead of the other way around.

Even in the case where you have discussed the parameters of an offer on the phone and suggested that you would respond favorably, you still have time to back out. Note that I said "suggested" that you would respond favorably. You learned in Chapter 8 that you should *never* utter the word "yes" to a job offer until you have it in writing and have reviewed it thoroughly. The most positive statement you should make after they have made you a definitive offer on the phone is one we provided earlier:

> *First, thanks for working with me on this offer; I really appreciate it. What you said sounds pretty much like what we've been discussing. If that's the case, then as I said earlier, I'm feeling very positive about this. Clearly, before I give you a definite yes, I'm going to need to see all this in writing, and if all is in order then I think we'll have a deal. Does that work for you?*

So, you have not said yes, the letter comes, and it is exactly what you wanted and had agreed to. Can you change your mind at this point? The answer is of course yes; you can change your mind right up to the point of signing the offer letter and technically even after signing it. But it is not good form. They have invested an awful lot of time in you and have presumably negotiated in good faith. Changing your mind at this stage will create ill will toward you from everyone involved, and they will remember it.

REASONS TO WALK AWAY

Here are some of the factors that could cause you to change your mind.

The company is not as solid or stable as you thought

You came to this conclusion after you did more due diligence on the company. You could have done this research before you sent in your résumé, but if you are sending dozens of applications, why research them if you are not even going to get an interview? The time to start is when you get a screening interview. But unless you are a financial sleuth, ascertaining the fiscal stability or longevity of a company may not be easy.

For a public company, one whose shares are traded on one of the big platforms like the New York Stock Exchange (NYSE) or NASDAQ, you can examine recent share price trends, resulting changes in market valuation, and significant corporate announcements on a variety of online platforms like Yahoo! Finance.[112] Warning signs might include a significant drop in share price, possibly associated with a company announcement of the failure of a major program or the cancellation of a corporate partnership. If the company is public, their required annual and quarterly US Security and Exchange Commission (SEC) documents referenced in the previous chapter provide a wealth of financial and corporate information. But these are full of boilerplate language. It can be difficult to discern the important information unless you are experienced at it.

If you have applied to a private company, the information available may be limited or nonexistent. Some early-stage companies may be operating in "stealth" mode and do not want any information about them to be made available to the public for a variety of reasons. Most, however, will post announcements on their website relating to corporate partnerships, fundraising successes, and key research advances. But extrapolating from such information to ascertain their likelihood of success is no more reliable than consulting a Ouija board. Everything a private company puts on the web is carefully written to convey a positive image.

The best source of information about the future of a company comes from insiders, including the people you meet with or who interview you. I am constantly amazed by the willingness of employees to share or hint at nonpublic information during interviews. No matter how much management admonishes them to keep their lips sealed with regard to company events, prospects, and failures, there will be those who either forget or ignore this and spill the beans.

So, do not be shy about asking the questions we listed in Chapter 6, as well as these:

"How would you say things are going here from a scientific perspective? Do things look promising?"

If you get an evasive answer to this, and you also learn that they have only twelve months of operating cash on hand, there may be a reckoning coming. You will need to decide for yourself what your risk tolerance is when joining a startup. See Appendix 1 at the end of this chapter for more insight into joining a startup company.

"I know this might sound weird, but if you were looking for a job today, would you still take your job here?"

While you would not ask this of anyone in a management role, you could ask someone who might be a prospective peer, but only do so in a one-on-one conversation.

"If you were in my position, is there anything you'd like to know about this place before signing on?"

The caveats for this question are the same as for the previous question.

"Because you are a private company, it was hard for me to get an idea of your finances. Could you give me the big picture of how you are fixed for money? When will you have to go back to investors to raise more?"

Ask that of anyone in a management or leadership role. There is no reason for them to dissemble or evade such a question unless their financial prospects are dim.

"Have there been any significant layoffs or turnover in senior staff or other levels? What was behind that?"

This can be asked of anyone. Layoffs per se are not necessarily red flags because programs and associated staff do get axed with some frequency in industry. Large or established companies terminate programs for many reasons including failure of a clinical trial, changes in the market or consumer demand, and seemingly random or inexplicable shifts in corporate priorities. For smaller companies, such events may be sentinels that they are running out of money and need to focus on programs they deem more likely than others to succeed. Significant changes in management or senior leadership, especially if they involve more than one person being let go or seem to occur with some regularity, may signal organizational instability or dysfunction.

"What would you say is your most promising program right now, in terms of likelihood of success? What do you think is the biggest risk to this program?"

You are looking for "loose lips." Watch their facial expression or body language when they try to answer this question. If the rank and file are skeptical of the company's programs and let slip that there are significant hurdles to success, this might be cause for concern.

You have concerns about their leadership

"What is the CEO like? What about the other senior leaders?"

This is an open-ended question that requires a substantive answer, as opposed to asking "Is the CEO nice?" or "Is she easy to work with?" which can be too easily answered with a reflexive "yes." Watch for telltale delays or vague generalities that may suggest the person does not want to or cannot give an honest answer. The working environment and culture of an organization are shaped more than most people realize by the management styles, personalities, and, at times, neuroses of senior leaders. Even if you do not expect ever to cross paths with the CEO, dysfunctional leadership at the top is contagious. The worst that can happen if you ask these questions is that they will defer, demur, or deny ("I don't know"). The best is that one of the less filtered of them will tell you something that gives you pause concerning the culture or environment of this place.

If you are applying for an entry-level position, be thoughtful about who you ask questions like these. Some will see it as a mark of sophistication, and others may find it inappropriate.

A few years ago, I applied for a job as the administrative head—basically the Chief Operating Officer—of a large private scientific research foundation. I had done a lot of reading about the organization, which revealed considerable internal strife and even disorganization. I had no problem with this and saw it as a challenge that I relished. They flew me in for an interview, which started with my meeting the president and then about four of their senior staff members, one by one. I quickly learned from each of them that the single biggest challenge they faced was that the chairman of their Board of Directors wielded immense power and influence. The chairman was a very wealthy and successful businessman who had played a pivotal role in the creation and funding of the foundation. I was told that he had a deep proprietary interest in its operations. In fact, he routinely overrode scientific and strategic decisions that the senior scientific staff made. Everyone was very open about expressing their frustration to me, and they clearly were hoping that if I came, I could help rein in this individual.

I was scheduled to meet with the chairman the next morning at 8:00 in the office he kept at the institute. I showed up at 7:50. The institute was empty at that hour save for his assistant, who ushered me into a waiting area outside his office. At 8:15 he arrived, walked right by me without acknowledging my presence, despite the fact that I was sitting right outside his office in an otherwise empty building, entered his office, leaving his door wide open, and proceeded to engage in one phone call after another, talking in a loud voice, for the next hour. At around 9:30 his assistant told me that he would see me. I met with him for about thirty minutes, during which time he asked me not a single question but rather explained exactly how he ran the organization and what his plans for the future were. I listened politely and was ushered out of his office when he was done talking. At this point my decision to withdraw my application had been made. While I relished the challenge, this job seemed more like an exercise in self-destruction.

I sent the president an email the next day respectfully withdrawing my application on the basis that the position wasn't a good fit. The president called me the next day to try to dissuade me, but he knew exactly why I was withdrawing without my having to spell it out in great detail, and he didn't really expect me to relent. We parted on good terms. I followed the organization's press announcements to see if they would fill this position. They did, one year later, with someone who lasted exactly six months. I figured I dodged a bullet.

The people you met were not friendly

This may sound odd, but it happens. You would think that if they are serious about you as a candidate they would be on their best behavior. When this happened to me it was typically because the people I was talking to were unhappy in their jobs or saw my candidacy as a threat to themselves. If it is just one or two people, maybe you can convince yourself you can live with it, unless that one person is someone you are going to be interacting with regularly, managing, or worse, managing you. If it is more than one person, you will be wondering whether it will be possible for you to get yourself out of bed every morning to go to work.

I was being interviewed for a senior scientific position at a small company of about thirty employees. I had met the CEO at an industry event where she told me that they needed someone to impose some order and structure in what was currently a rather dysfunctional organization. She invited me to visit the company, and we set up a date to do this. On my arrival the CEO spent about an hour preparing me to meet with three of the senior scientists in the company. She described one of them, Andrew, as being the current de facto head of their scientific operations. She said that Andrew was very dogmatic and opinionated but was the most technically skilled scientist in the company. She said that he was likely to be threatened by my presence and might be somewhat aggressive when we met. She got all of that right.

I had four meetings scheduled with her staff, each lasting thirty minutes. She escorted me to a conference room and placed a small clock on the table, telling me that I could use the clock to know when it was time for the next person to come in. Andrew was my first meeting. He walked into the room, didn't introduce himself, didn't offer to shake hands, sat down across the table from me, and turned the small clock so that it faced him and so that I couldn't see it. At this point I think I knew this was never going to work out. I figured right away that the CEO wasn't going to do anything that would threaten this key scientist. But, on an impulse, I reached over and turned the clock halfway so that we could both see it. Andrew glared at me, and we proceeded with the interview. I relayed all this to the CEO after my meetings, and she said she would debrief Andrew. I knew that this was not going to work out but decided to wait to hear from her rather than withdrawing. Not surprisingly, I didn't get the job, which was just as well.

You heard concerning stories from some of their staff

In Chapter 6, we listed dozens of questions for you to ask during your interview. Some open-ended questions that could elicit what you are looking for include the following:

- If you were coming to this department, what would you like to know?
- Are there currently ongoing tensions regarding social or political matters that I should know about?
- How would you describe the culture here?

Academics in particular may answer such questions with shockingly brutal honesty, making you wonder whether they are intentionally trying to talk you out of coming.

You met the hiring manager and cannot imagine working for them

Does this even require further comment? In many startups early hires get put into management roles simply because they were there. The company does not have the time or resources to recruit more experienced people, so you are stuck with the person who just got their Ph.D. in solid state physics as your boss.

Here's a "pro tip": If the company hiring you does not have a human resources department, or worse, does not even have a human resources person, be on the alert. Without HR support, staff may not be getting the support they need to navigate interpersonal or managerial challenges. Managers and supervisors with no training in managing or supervising are common in science organizations and almost universal in startups. Even a single HR person can serve as a buffer between inexperienced managers and shell-shocked staff. In larger or more established organizations the presence of truly bad managers is less common because there are built-in oversight mechanisms by which such behaviors are uncovered and corrected.

You have concerns that you may not get the kind of mentoring you want or need

This is a special case of the above situation where the position you are applying for is largely or even partly one in which you expect to receive training. The prime example is that of a postdoctoral position. Some postdoc positions are explicitly referred to as "postdoctoral training" positions, possibly related to the agency that funds the position using that term. Whether or not the word *training* is mentioned in the position description, nearly all postdoc positions are implicitly training experiences. The postdoc is expecting to learn a new field or technique to expand their skill set and, in the best cases, to develop a degree of scientific maturity and independence. Even for positions advertised as "research scientist," in an academic setting, the best will provide the applicant with valuable training and mentoring experiences that will help them not only in that job but in advancing their career.

If you are applying for a postdoc position, you should make it a point to speak with others who are currently or were previously postdocs in that lab. Ask plenty of questions about their experiences with the PI. Avoid overly general questions like "Is X a good mentor?" because they are too easy to answer with a yes or no. Better questions are ones like these:

- Does X meet with you regularly? How often?
- Do they give you helpful feedback and guidance about your work?
- Are they helpful in providing you with career advice?
- Have they introduced you to others in the field?
- Have they given you opportunities to attend and/or present at scientific meetings?
- Have they helped you prepare an Individual Development Plan?
- How much independence do you have in your work?
- Do you have the opportunity to spend part of your time on an independent or semi-independent project that you could take with you when you leave?

If the answers you get to these questions suggest that this PI is not going to provide you with the kind of mentoring you want, then perhaps you should look elsewhere.

You got bad vibes about their culture

This sounds like a distressingly nonscientific criterion. But what if you get the feeling that there is something about the people or the job that just does not feel right even though you cannot put your finger on it? Is that enough for you to move on? It all depends on the source of your discomfort. Try explaining your bad vibes to a few friends and see how they react. If their reaction is that you are making mountains out of molehills, misinterpreting things you have seen or heard, or viewing your interview experience through the lens of your own neuroses and insecurities, then maybe you should put your bad vibes aside. If on the other hand they agree that you have picked up on some fundamental issues that would also give them pause, then your friends have earned your friendship.

You decided you do not want to live there

You should have figured this out before they interviewed you, but maybe you are one of those people who thinks "Despite my deep existential need to live in a bucolic natural environment where I can keep chickens in my yard, I think I'll be able to adapt to living in a two hundred-square-foot studio apartment in a large metropolitan area where the closest grass is a half mile away." Or the converse: "I really cherish having access to a vibrant arts and entertainment community set in a bustling urban setting with artisanal coffee shops on every corner and where I can rent a bike or scooter, so I don't need to own a car, but central Iowa and the great plains do have their charms." Think twice before agreeing to be interviewed if either of these applies to you.

You cannot reach an agreement on key terms of your contract

This is the most obvious and easiest reason to justify turning down an offer. If you have gone through the exercise in Chapter 9 where you categorized the various items on your wish list in terms of ideal, acceptable, and bottom line, and some key elements are below your bottom line even after several negotiations, then this job is not for you. In the next chapter I will show you how to decline the offer in this situation.

You do not see growth opportunities there

Opportunities for advancement or growth are key attractive elements for many. You may not want to be trapped in what seems like a dead-end job doing the same things year after year. Here are some signs that advancement opportunities may be lacking: employees told you so when you asked them; the organization has a flat structure with few or no middle managers; and the company has shown a predilection for hiring from the outside as opposed to promoting from within. A way to find out if the last practice is prevalent is to ask people you meet with how long they have been in their current roles and what they were doing before that. If they were promoted from within, you will hear about it.

They do not have the resources that you need

Even if you think they are going to give you everything you asked for in terms of salary, space, startup funds, etc., and they are terrific people who wined and dined you at a fancy restaurant, you may still conclude that this is not going to work. But it could take some time and digging before you figure that out. For an academic job, Chapter 7 lists the questions to ask the faculty, the chair, or the dean to help you ascertain whether this place will be able to support you and your proposed program. Other key factors include the following:

- Lack of access to a pool of high-quality graduate students.
- Absence of specific expensive or specialized equipment that you will need but could not acquire on your own.
- Absence of important core or support services that are or will be at some point essential to your success (animal facility, core cell culture, proteomics, imaging, mass spec, computing, bioinformatics, sequencing, structural biology, NMR, fabrication facilities).
- Absence of others working in a research area even remotely close to yours, either in the department or in the entire institution. Be careful about this one though. In my experience, scientists have vivid fantasies about having random and immensely satisfying conversations with like-minded peers over coffee or lunch, and in the corridor outside their office. The reality is that 95% of your time will be spent in your office and lab talking to people who work for you. The remaining 5% will be on the phone, Zoom, or email communicating with collaborators. Granted, it is nice to bump into people, but if you are honest with yourself, you will probably admit that a) it does not happen as often as you think, and b) when it does happen, you end up talking about sports or politics, which you could as easily do with a social anthropologist as you could with a molecular geneticist.

You got a better offer

This happens more frequently than you might think. If you have applied and interviewed for numerous positions, and you are a good enough candidate to get one offer, chances are you are going to get others, provided you applied to enough places.[1] Each organization operates on its own clock. So, it is likely that one will take longer to make up its mind than another, or their offer will come only after their first choice turned them down. Even if you have gone a long way down the negotiating path with the first organization, you can and should consider the new offer. If you really want to go to the first place, but the second place makes you a better offer, you can use that to negotiate better terms with the first place. The first place may ask to see the written offer letter from the second place with the better terms spelled out, and you should show it to them. You are negotiating your professional future, and you have an obligation to yourself and your science to take the position that offers you the best chance of success.

It is easy in the frenzy of applying, interviewing, and negotiating an offer to lose track of the fact that the final decision in taking a job is yours and yours alone. Ending up in a job that is not really what you want to be doing, in a place you do not want to be, and with people you are not going to look forward to working with can be a lot worse than being on the job market for another six months.

CHAPTER 14 SUMMARY

Is this job for you?

- Recognize when a job is not the right fit for you before it is too late.

Warning signs

- Be on the alert for warning signs that can cause you to reconsider joining a lab or organization, including company financial instability, poor leadership or mentoring quality, and unfriendly/hostile staff.

Continuous evaluation

- Constantly use new information that you gather throughout the interview process to update your assessment of whether this job is for you. Make decisions based on your career goals instead of feeling obligated to continue with the process.

Withdrawal strategy

- Know how to diplomatically withdraw from applications while being tactfully vague about specific reasons.

Startup reality check

- Know the characteristics of startups so you can decide whether that is a good fit for you.

Have patience

- Taking the wrong job can be worse than remaining on the job market a bit longer.

Quality over speed

- Prioritize finding the right fit over accepting the first available offer.

APPENDIX 1

Is a startup right for you?

Being an early hire in a startup is one of the most exhilarating experiences you can have. But it can also lead to disillusionment and disappointment if you do not recognize the natural trajectory of new companies as they grow. With growth comes changing roles as new and sometimes discordant voices are added. Many early hires experience a sense of loss for the good old days. Preparing for and navigating these changes can help you maintain your commitment and value to the organization. A typical trajectory we often see looks like this:

1. You were one of the first hires and probably found yourself interacting with the CEO and other early hires and doing a little bit of everything, even if your position was a scientific or technical one. These were exciting times with lots of camaraderie and long hours.
2. During those early days you likely played a key role in interviewing and selecting new hires, discussing strategy, and even meeting with potential investors. You felt like a member of the leadership team with a tangible stake in the growth of the company.
3. Gradually, as more staff were hired, you started doing what you were hired to do. Perhaps you were put in charge of a team focused on a specific project or providing an important technical service. Here is where things began to change for you. As the company grew, and new managers and technical contributors were hired, you found yourself being left out of meetings and not having a voice in decisions that you were previously part of. Even though you understood the necessity behind these changes, you did not like them.
4. Maybe you became resentful of these changes in your role and involvement. If you thought about it, you might have recognized the origin of your resentment—a desire for things to revert to the way they were. Recognizing this may have helped you manage your resentment, which if unacknowledged might have resulted in your acting it out. While being disappointed is natural, acting disappointed is not professional. Ways in which your feelings might have been manifested as actions include withdrawing from group discussions, acting resentfully towards those in leadership, or being hypercritical of new hires or peers in order to regain your old influence. These behaviors would have been to your detriment, resulting in your no longer being seen as a team player and further distancing you from the centers of power and decision-making where you longed to be.
5. What did you do about it? Did you:

    A. Accept that roles change with growth and that you will inevitably feel a sense of loss and a lack of control over these changes?

B. Recalibrate your expectations as your role changed so that you could feel as committed as you were in the early days?

C. Learn new skills that enabled you to contribute in ways that the company needs now rather than in ways they no longer need?

D. Never stop being a supporter of the team and your colleagues?

If you did these things, it is likely that you are as valuable and connected now as you were then. If you did not, you are probably wondering "What happened to my beautiful tech startup?"

15 Getting to No

Understanding Rejections and Declining Offers

- **When they are sorry to inform you**
  - There was another candidate who was a better fit
  - Dysfunctional search
  - No one got it
  - It was a pretend search
  - You made a big mistake during your interview
- **When you are sorry to inform them**
- **Chapter 15 summary**

Most of you will end up applying for dozens or even hundreds of jobs over the course of your career. In many or even most cases you will not get offered the job you want, and in other cases you may not want the job you get offered. Every time you do not get a job you will obsess over why you did not get it, and, in every case, you will have precisely zero information why. You will then make up your own reasons why you did not get the job, and in almost every instance you will be wrong. Here are some of the most common reasons why you might not have gotten the job.

WHEN THEY ARE SORRY TO INFORM YOU

There was another candidate who was a better fit

This is the most common reason, and it makes perfect sense that it would be. The hiring manager or committee or entire faculty had to make a choice between the finalists, and they chose the one who they thought would best meet their needs and search criteria. This does not mean that you were a "bad" candidate, only that they thought the one they chose had a mix of skills that best matched this job. We use "thought" to emphasize that this is in part a subjective decision made as much on the basis of "gut" feelings as anything else.

Dysfunctional search

This is more common than you think. In this case, the position description was cobbled together by warring factions within the organization, each with fundamentally different conceptions of what kind of candidate they should hire. The position description that resulted was either incredibly vague, overly specific and descriptive of no candidate who could possibly exist, or agnostic, as in "open to candidates in any discipline or field." The result is that when it came to candidate selection and interviews, everyone was looking for a set of skills that were different from, and incompatible with, what others were looking for. The good news for you if you did not get the job is that dysfunctional search processes are hallmarks of dysfunctional departments or organizations.

No one got it

Just because you did not get hired does not mean someone else did. In some cases, possibly because of a dysfunctional search, numerous people were interviewed, but the hiring manager or committee were unable to find a candidate who met their search criteria. They decided that they can wait and try again in six months or that they can live without filling this position.

It was a pretend search

A pretend search is one that is conducted as a formality to enable them to hire a candidate who they have previously identified but who cannot be hired without their having done a formal search. In many cases this is an internal candidate—someone already in the organization who everyone knows. Especially for high-level positions, organizations will undertake a remarkably thorough external search, including screening interviews, in-person interviews, and even reference checks to ensure that the pretend process looks credible. A pretend search is not to be confused with a pretend advertisement. These are posts for nonexistent positions used for a variety of dubious or illicit purposes, including by "spec" recruiters to collect your résumé. These have become alarmingly common but typically do not lure you any further than submitting your résumé.[113]

You made a big mistake during your interview

This is always the number one fear in a candidate's mind when they do not get the job, but it is rarely the real reason. Spilling your wine at dinner, momentarily forgetting the CEO's name, or showing a slide with a mislabeled *x*-axis will be embarrassing in the moment, but chances are they will forget about it in an hour, did not care in the first place, or did not even notice. If they have selected you from dozens or hundreds of applicants, interviewed you in person, and spent untold hours on your candidacy, it is unlikely that one incident is going to end up being the deciding factor.

WHEN YOU ARE SORRY TO INFORM THEM

Just as there could be many reasons why you did not get a job offer, there are also many reasons why you might decide to drop out of consideration or not take a job that is offered. In the previous chapter we reviewed some of these reasons. Whatever the reason for your decision to drop out, the sooner you make it the better. There is no shame in getting to a certain stage of the process and then backing out. It could happen as soon as your screening interview when you learn something about the job that was not clear or even mentioned in the job posting. If that happens, email the person you met with, thank them for their time, and tell them you are going to withdraw your application for the position. You do not need to go into a lot of detail.

If you are declining an offer or withdrawing from the process because you disliked the culture or the leadership or the people you met, it is not a great idea to tell them that.

Taking the example of the situation in the previous chapter of the senior executive who dropped out after meeting the Board chair, you might send the following letter:

> *Dear Dr. X,*
>
> *Thank you so much for taking the time to meet with me yesterday. It was very informative. I am impressed by your company and by the remarkable progress that it has made over the past three years. However, after thinking more about this position, I have decided that it really isn't the best fit for me. So, I'm going to withdraw my application at this point. I wish you and your company the best and hope our paths cross again in the future.*
>
> *Best wishes,*
> *Dr. Y*

Points about this email to note include the following:

1. You thanked them for their time.
2. You said nice things about their company.
3. You did not say anything bad or pejorative about them or the people you met.

The reasons for not being more specific should be obvious, but let us spell them out. First, you never know when or under what circumstances you will meet these people again. You may apply to the same place five years from now, or someone who interviewed you will move to a different company or university where they will be interviewing you next year. Keep your eye on the long game—people in your profession will be around for a long time. There is no value in alienating them.

Second, if you make up a reason, such as you tell them it is about salary or your startup package, you need to be prepared for them to take you at your word and make you the offer you want. In that case you are stuck. If you say no again, you will be seen as a manipulator or dissembler. The take-home message is not to make up issues that they could address in a counteroffer. If you give specific reasons why you are turning them down and they address those reasons, you are in a very awkward position.

If their terms do not work for you and declining is something you can live with, send them a nice email thanking them for their time and interest and express sincere appreciation for their efforts. In declining an industry offer, do not spell out your specific reasons since if you do it will look like you are still negotiating. The opposite holds if you are declining an academic job offer because you have decided to take another offer. Letting the declined institution know that this is the reason is perfectly acceptable and even recommended. In academia, they are going to find out one way or another. Most academic hiring committees understand that good candidates will have other offers and will not hold a negative decision on their offer against you. An exception might be if they think you were "playing" them along only to use their offer to increase the terms of another offer. They will get over it eventually.

As we have seen in earlier chapters, interviewing for an academic job is a much more involved and time-consuming process—for both you, the candidate, and the hiring institution—than for one in industry. If you need to decline an academic offer for whatever reason, you should let the people at the declined institution know how appreciative you are of the effort and time they put into the process. This is not just the polite thing to do; it is also the politically beneficial thing to do. The people who interviewed you are likely to work in your field, and you will be interacting with them professionally in the future. You do not want to leave them feeling unappreciated. In some cases, the institution you are declining may deserve more than a letter or email, perhaps because they showed an especially keen interest in you and your work or you felt a personal connection with members of the search committee. In those cases, you should speak directly with the chair of the department or the hiring committee to thank them and to express your appreciation for their offer.

Whether you are declining an academic or industry offer, your "no" is not a tactic (even though they may interpret it as such), it is a decision. Nonetheless, if you do decline, be prepared for the possibility that they will then reconsider and meet your terms after being pushed to the brink. That is their decision and theirs alone; you will then have earned a reputation as a skilled negotiator.

CHAPTER 15 SUMMARY

Most job rejections have nothing to do with your performance

- Common reasons include finding a better-fit candidate, dysfunctional search processes, failed searches where no one gets hired, or "pretend searches" for predetermined internal candidates. Making a big mistake during interviews is rarely the actual deciding factor.

Failed searches often indicate dysfunctional organizations

- If no one gets hired after extensive interviews, it may signal warring factions with incompatible visions, overly vague or impossible job descriptions, or organizational dysfunction. Consider it a bullet dodged rather than a personal failure.

Be prepared for pretend searches where an internal candidate was preselected

- Some organizations conduct thorough external searches as a formality to hire someone they have already chosen internally. These fake processes can involve full interviews and reference checks to appear credible.

When declining offers, it is best not to give specific reasons

- Do not fabricate negotiable issues when your real reasons are not fixable. If you enumerate made-up reasons, they might come back with solutions. Stick with this: "After careful consideration, I don't think this is a good fit for me professionally."

References, Notes, and Resources

1. Fernandes JD, Sarabipour S, Smith CT, Niemi NM, Jadavji NM, Kozik AJ, Holehouse AS, Pejaver V, Symmons O, Bisson Filho AW, Haage A. 2020. Research culture: A survey-based analysis of the academic job market. *eLife* **9**: e54097.
2. Saltzman J. 2025. Vaccine firm finds few protections from backlash. *Boston Globe*, updated June 15. https://mobileapp.bostonglobe.com/06152025_dac92afc-494e-11f0-96ef-41e9351edff7/content.html.
3. Cohen CM, Cohen SL. 2018. *Lab dynamics: Management and leadership skills for scientists*, 3rd ed. Cold Spring Harbor Laboratory Press, Cold Spring Harbor, NY.
4. Skrentny JD. 2023. *Wasted education: How we fail our graduates in science, technology, engineering and math*. University of Chicago Press, Chicago.
5. Lahart J. 2025. There is now clearer evidence AI is wrecking young Americans' job prospects. *Wall Street Journal*, August 26. https://www.wsj.com/economy/jobs/ai-entry-level-job-impact-5c687c84?st=BBL4g8&reflink=desktopwebshare_permalink.
6. GlobalData Healthcare. 2025. Biopharma venture financing declines 20.2% YOY in Q1 2025 amid persistent investor caution. *Pharmaceutical Technology*, April 30. https://www.pharmaceutical-technology.com/analyst-comment/biopharma-venture-financing-q1-2025/.
7. Witze A. 2025. 75% of US scientists who answered *Nature* poll consider leaving. *Nature*, March 27. https://www.nature.com/articles/d41586-025-00938-y.
8. Pain E. 2025. How do research careers compare across countries? New global database has some answers. *Science*, June 30. https://www.science.org/content/article/how-do-research-careers-compare-across-countries-new-global-database-has-some-answers.
9. The status of non-tenure-track faculty. American Association of University Professors. https://www.aaup.org/reports-publications/aaup-policies-reports/topical-reports/status-non-tenure-track-faculty [accessed December 16, 2025].
10. Being a Ph.D. student shouldn't be bad for your health. 2019. *Nature*, May 15.
11. Busch C, Wiesenthal N, Gin L, Cooper K. 2024. Behind the graduate mental health crisis in science. *Nature Biotechnology* **42**: 1749–1753.
12. Cohen C. 2020. Management skills for scientists: A new institutional responsibility. Science Management Associates, January 2. https://www.sciencema.com/news/2020/1/2/management-skills-for-scientists-a-new-institutional-responsibility.
13. Citrin-Sadadi A. 2025. AI is wrecking an already fragile job market for college graduates. *Wall Street Journal*, July 29. https://www.wsj.com/lifestyle/careers/ai-entry-level-jobs-graduates-b224d624.
14. Woolston C. 2024. How to find your place in science through an industry postdoc. *Nature*, December 19. https://www.nature.com/articles/d41586-024-04169-5.
15. Kahn S, Ginther D. 2017. The impact of postdoctoral training on early careers in biomedicine. *Nature Biotechnology* **35**: 90–94.

16. Udesky L. 2025. Survey of US postdocs finds threefold increase in job losses. *Nature*, July 10. https://www.nature.com/articles/d41586-025-02140-6.

17. Smith R. 2025. AI is forcing the return of the in-person job interview. *Wall Street Journal*, August 12. https://www.wsj.com/lifestyle/careers/ai-job-interview-virtual-in-person-305f9fd0.

18. To learn more, read Chapter 11, "Science Inc., making a smooth transition to industry," Cohen and Cohen, *Lab dynamics*.

19. We will use in this book terms that designate the type of educational institutions from the Carnegie Foundation. The 2025 Research Activity Designations include the following categories:
    - Research 1(R1): Very High Spending and Doctorate Production, now defined by a clear threshold of $50 million in total research spending and 70 research doctorates awarded annually. In 2025, 187 institutions were given a designation of R1.
    - Research 2 (R2): High Spending and Doctorate Production, which continues the previous requirement of $5 million in research spending and 20 research doctorates awarded annually. This year, 139 institutions received a designation of R2.
    - Research Colleges and Universities (RCU), a new designation that identifies research happening at colleges and universities that historically have not been recognized for their research activity, including institutions that do not offer many or any doctoral degrees. This new designation includes any non-R1 or R2 institution that spends more than $2.5 million on research annually, accounting for 216 institutions.

20. Cappelli P. 2019. Your approach to hiring is all wrong. *Harvard Business Review*, May–June.

21. Granovetter M. 1973. The strength of weak ties. *American Journal of Sociology* **78**: 1360–1380.

22. Chakraborty T, Mekkawy H, Cloete K, Mahla R, Peng D, Perez A, Rouabah N, Gomes J, Drallou F, Abdalrahman T, et al. 2025. Supporting scientists who study and work abroad. *Science*, July 3. https://www.science.org/doi/10.1126/science.aea0626.

23. Streeter J. 2014. Networking in academia. *EMBO Reports* **15**: 1109–1112.

24. Weber L. 2025. Millions of résumés never make it past the bots: One man is trying to find out why. *Wall Street Journal*, July 22. https://www.wsj.com/lifestyle/careers/ai-resume-screening-hiring-676a4701.

25. Employers are buried in A.I.-generated résumés. 2025. *New York Times*, June 21. https://www.nytimes.com/2025/06/21/business/dealbook/ai-job-applications.html.

26. Moore, T. 2014. Why women don't apply for jobs unless they're 100% qualified. *Harvard Business Review*, August 25. https://hbr.org/2014/08/why-women-dont-apply-for-jobs-unless-theyre-100-qualified.

27. Coffman K, Collis M, Kulkarni L. 2023. Whether to apply. *Management Science* **70**: 4649–4669. https://pubsonline.informs.org/doi/10.1287/mnsc.2023.4907.

28. Lerner H. 2015. *The confidence myth: Why women undervalue their skills, and how to get over it*. Berrett-Koehler, Oakland, CA.

29. How to write effective resumes and CVs for doctoral students and postdocs. https://imagine.jhu.edu/resources/how-to-write-effective-resumes-and-cvs-for-doctoral-students-and-postdocs/ [accessed December 30, 2025].

30. Cai H, Wu L, Shi Y, Gu R, Sedikides C. 2016. Self-enhancement among Westerners and Easterners: A cultural neuroscience approach. *Social Cognitive and Affective Neuroscience* **11**: 1569–1578.

31. Boothroyd J. 2019. How to prepare a compelling research statement for R1 faculty applications. https://www.youtube.com/watch?v=kYZ8HP6f9Jc.

32. Survey of earned doctorates. National Science Foundation. https://ncses.nsf.gov/pubs/nsf24300/report/about-this-report [accessed December 30, 2025].

33. Education, training, and academic R&D. National Science Foundation. https://ncses.nsf.gov/pubs/nsb20213/education-training-and-academic-r-d [accessed December 30, 2025].

34. Faculty application: Teaching statement. MIT EECS Communication Lab. https://mitcommlab.mit.edu/eecs/commkit/faculty-application-teaching-statement/ [accessed December 30, 2025].
35. Margaria C. 2016. Teaching statement. https://www.bu.edu/econ/files/2017/10/teaching_statements.pdf.
36. Faculty application: Teaching statement. MIT EECS Communication Lab. https://mitcommlab.mit.edu/eecs/commkit/faculty-application-teaching-statement/ [accessed December 30, 2025].
37. Kelskey K. 2015. *The professor is in*. Three Rivers Press, New York.
38. Koehler J. 2025. Meta is going to let job candidates use AI during coding tests. *Wired*, July 29. https://www.wired.com/story/meta-ai-job-interview-coding/.
39. Hanson W. 2025. *Just good manners: A quintessential guide to courtesy, charm, grace, and decorum*. Gallery Books, New York.
40. Kahneman D. 2011. *Thinking fast and thinking slow*. Farrar, Straus and Giroux, New York.
41. Tye K. 2019. Faculty search committees: Once you get to the interview stage, what is the single most important factor in candidate selection? *Twitter*, September 18. https://twitter.com/kaymtye/status/1174311929789071360.
42. Sinek S. 2009. Start with why—how great leaders inspire action. https://www.youtube.com/watch?v=u4ZoJKF_VuA.
43. Mackiewicz J. 2006. Audience perceptions of fonts in projected PowerPoint text slides. IEEE International, October 23–25. https://doi.org/10.1109/IPCC.2006.320391.
44. See this excellent guide to answering questions during a talk: http://career.ucsf.edu/QandA.
45. Mahley R. 2020. Preparing a chalk talk for a faculty position. Gladstone Institutes. https://www.youtube.com/watch?v=03VVfk7edCc.
46. Snapp EL. 2019. How to design a chalk talk—the million dollar sales pitch. *Molecular Biology of the Cell* **30**: https://www.molbiolcell.org/doi/10.1091/mbc.E19-01-0007.
47. Smith MK, Wenderoth MP, Tyler M. 2013. The teaching demonstration: What faculty expect and how to prepare for this aspect of the job interview. *CBE Life Sciences Education* **12**: 12–18.
48. The Chronicle's salary data library. The Chronicle of Higher Education. https://data.chronicle.com/ [accessed December 30, 2025].
49. Faculty salaries. Inside Higher Ed. https://www.insidehighered.com/content/faculty-salaries [accessed December 30, 2025].
50. The annual report on the economic status of the profession, 2022–23. 2023. American Association of University Professors. https://www.aaup.org/report/annual-report-economic-status-profession-2022-23.
51. Search salary now. H1B Salary Database 2025. https://h1bdata.info/index.php [accessed December 30, 2025].
52. Homepage. Indeed. https://www.indeed.com/?from=gnav-acme–acme-webapp [accessed December 30, 2025].
53. Salary guide 2026. Randstad. https://tinyurl.com/bdhums5c [accessed December 30, 2025].
54. Guide salaries. Glassdoor. https://www.glassdoor.com/Salaries/guide-salary-SRCH_KO0,5.htm [accessed December 30, 2025].
55. Overview of BLS wage data by area and occupation. U.S. Bureau of Labor Statistics. https://www.bls.gov/bls/blswage.htm [accessed December 30, 2025].
56. Tools for determining salary benchmarks. Yale School of the Environment, Office of Career and Professional Development. https://careers.environment.yale.edu/resources/tools-for-determining-salary-benchmarks/ [accessed December 30, 2025].

57. Faculty job search. Future PI Slack. https://futurepislack.wordpress.com/faculty-job-search/ [accessed December 30, 2025].

58. Biology. Google Sheet. https://docs.google.com/spreadsheets/d/1Oi0B01kb5S9bleY1XuZBE0kzzercm7Y2sCZUJrp8XZs/edit?gid=1711743605#gid=1711743605 [accessed December 30, 2025].

59. University of California (UC) employee pay. University of California. https://ucannualwage.ucop.edu/wage/ [accessed December 30, 2025].

60. Ury R. 1993. *Getting past no*. Bantam, New York. And for applications to a lab setting, see Cohen and Cohen, *Lab dynamics*.

61. Purdy JM, Nye P. 2000. The impact of communication media on negotiation outcomes. *International Journal of Conflict Management* **11**: 162–187.

62. Ebner N. 2017. Negotiation via email. In *The negotiator's desk reference* (ed. Honeyman C, Schneider AK). DRI Press, St. Paul, MN.

63. Fry R, Aragão C. 2025. Gender pay gap in U.S. has narrowed slightly over 2 decades. Pew Research Center, March 4. https://www.pewresearch.org/short-reads/2025/03/04/gender-pay-gap-in-us-has-narrowed-slightly-over-2-decades/.

64. The simple truth about the gender pay gap. American Association of University Women. https://www.aauw.org/resources/research/simple-truth/ [accessed December 30, 2025].

65. Höfrová A, Moore A, Small MA, Rosopa PJ, Payne KS, Rymešová P. 2024. Gender differences in faculty experience with start-up packages: A case study from a public university in the southeastern U.S. *Journal on Efficiency and Responsibility in Education and Science* **17**: 130–139.

66. Zernike K. 2023. *The exceptions: Nancy Hopkins, MIT, and the fight for women in science*. Scribner, New York.

67. Amanatullah ET, Morris MW. 2010. Negotiating gender roles: Gender differences in assertive negotiating are mediated by women's fear of backlash and attenuated when negotiating on behalf of others. *Journal of Personality and Social Psychology* **98**: 256–267.

68. Bowles HR, Babcock L, Lai L. 2007. Social incentives for gender differences in the propensity to initiate negotiations: Sometimes it does hurt to ask. *Organizational Behavior and Human Decision Processes* **103**: 84–103.

69. Heim P, Hughes T, Golant S. 2005. *Hardball for women: Winning at the game of business*. Plume, New York.

70. This scheme is based on one developed by Bill Lindstadt at the UCSF Careers Office, which is a source of much excellent guidance. See also Negotiating Your Faculty Compensation and Startup Package at https://career.ucsf.edu/events/pac-up/negotiating-your-faculty-job-offer and https://career.ucsf.edu/sites/g/files/tkssra2771/f/What%20Comes%20Next%20-%20Negotiating%20for%20a%20Faculty%20Position%20-%20November%202023.pdf.

71. Farris EM, Key EM, Sumner JL. 2023. "Wow, I didn't know these options existed": Understanding tenure-track start-up packages. *PS: Political Science & Politics* **56**: https://doi.org/10.1017/S1049096522000993.

72. Total salary range for Cornell University Assistant Professor. Glassdoor. https://www.glassdoor.com/Salary/Cornell-University-Assistant-Professor-Salaries-E3732_D_KO19,38.htm [accessed December 30, 2025].

73. Jaschik S. 2016. What you teach is what you earn. Inside Higher Ed, March 27. https://www.insidehighered.com/news/2016/03/28/study-finds-continued-large-gaps-faculty-salaries-based-discipline.

74. O'Leary B, Zahneis M. 2025. The faculty salary divide. The Chronicle of Higher Education, May 23. https://www.chronicle.com/article/how-faculty-pay-and-tenure-can-change-depending-on-academic-discipline.

75. Faculty: 4-digit multi-discipline report. University of Montana, April 3, 2025. https://www.umt.edu/provost/faculty/marketadjustment/faculty_4-digit_multi-discipline_report_2024-25.pdf [accessed December 31, 2025].

76. Kim N, Park H-J. 2017. Making the most of the first-offer advantage: Pre-offer conversation and negotiation outcomes. *Negotiation Journal* **33**: 153–170.
77. Tversky A, Kahneman D. 1974. Judgement under uncertainty: Heuristics and biases. *Science* **185**: 1124–1131.
78. Boss JM, Eckert SH. 2004. Academic scientists at work: Giving it 110%. *Science*, February 13. https://www.science.org/content/article/academic-scientists-work-giving-it-110.
79. Corso J. 2024. How to negotiate your first faculty job offer. Medium, February 8. https://medium.com/@jasoncorso/how-to-negotiate-your-first-faculty-job-offer-852beed1dd73.
80. Caromile LA, Segarra VA. 2024. Negotiating an academic start-up package and job offer as an incoming tenure-track professor in the life sciences. *BMC Proceedings* **18**: 29.
81. U.S. Bureau of Labor Statistics. https://www.bls.gov/ [accessed December 30, 2025].
82. Academic lab space. Tenure, She Wrote, March 12. https://tenureshewrote.wordpress.com/2015/03/12/academic-lab-space/.
83. Macdonald JE. 2019. Negotiating details in the academic letter of offer. *Chemistry of Materials* **31**: https://pubs.acs.org/doi/full/10.1021/acs.chemmater.9b04051.
84. McNeil L, Sher M. 1999. The dual-career-couple problem. *Physics Today* **52**: 32–37.
85. Schiebinger LL, Henderson AD, Gilmartin SK. 2008. *Dual-career academic couples: What universities need to know*. Stanford University, Michelle R. Clayman Institute for Gender Research. https://hps.stanford.edu/sites/hps/files/media/file/dualcareerfinal.pdf.
86. Schussler EE. 2025. Yours, mine, sometimes ours: Uncovering the experiences of biologists searching for academic jobs together. *CBE Life Sciences Education* **24**: ar22.
87. Wolf-Wendel LB, Twombly S, Rice S. 2003. *The two-body problem: Dual-career-couple hiring practices in higher education*. Johns Hopkins University Press, Baltimore.
88. For example, here is the University of North Carolina's policy: https://facultyaffairs.unc.edu/policies-and-procedures/faculty-recruitment/faculty-spousal-and-partner-hiring-assistance-program/.
89. Monahan T, Waltz M, Erol M, Fisher JA. 2024. Partner hire scorecard: A ranking of the partner-friendly status of U.S. universities. https://partnerhirescorecard.org/reports [accessed December 31, 2025].
90. Allen J, Smith JL, Ransdell LB. 2019. Missing or seizing the opportunity? The effect of an opportunity hire on job offers to science faculty candidates. *Equality, Diversity and Inclusion: An International Journal* **38**: 160–177.
91. Colby G. 2023. Data snapshot: Tenure and contingency in US higher education, Fall 2023. https://www.aaup.org/academe/issues/spring-2025/data-snapshot-tenure-and-contingency-us-higher-education-fall-2023.
92. Tower, C. 2001. Negotiating the non-tenure track. The Chronicle of Higher Education, July 6. https://www.chronicle.com/article/negotiating-the-non-tenure-track/?sra=true.
93. The status of non-tenure-track faculty. American Association of University Professors. https://www.aaup.org/reports-publications/aaup-policies-reports/topical-reports/status-non-tenure-track-faculty [accessed December 16, 2025].
94. A portrait of part-time faculty members. 2012. Coalition on the Academic Workforce. http://www.academicworkforce.org/CAW_portrait_2012.pdf [accessed December 31, 2025].
95. Flannery ME. 2024. 5 things to know about faculty pay today. NEA Today, April 30. https://www.nea.org/nea-today/all-news-articles/5-things-know-about-faculty-pay-today.
96. Here's what the average bonus looks like. 2025. Northwestern Mutual, February 14. https://www.northwesternmutual.com/life-and-money/average-bonus-percentage/.
97. Huston T. 2021. *Let's talk: Make effective feedback your superpower*. Portfolio, New York.

98. Adler L. 2007. *Hire with your head*, 3rd ed. Wiley, Hoboken, NJ.

99. Heifitz RA. 1994. *Leadership without easy answers*. Belknap Press, Cambridge, MA.

100. Bento F, Tagliabue M, Lorenzo F. Organizational silos: A scoping review informed by a behavioral perspective on systems and networks. *Societies* **10**: 56.

101. De Waal A, Weaver M, Day T, van der Heijden B. 2019. Silo-busting: Overcoming the greatest threat to organizational performance. *Sustainability* **11**: 6860.

102. Glen P. 2002. *Leading geeks*. Jossey-Bass, San Francisco.

103. Bennis W. 1994. *On becoming a leader*. Addison-Wesley, Boston.

104. Katzenbach JR, Smith DK. 1999. *The wisdom of teams*. Harper Business, New York.

105. Heifetz RA, Linsky M, Grashow A. 2009. *The practice of adaptive leadership*. Harvard Business Review Press, Boston.

106. Katzenbach JR, Khan Z. 2010. *Leading outside the lines*. Booz and Co., Washington, DC.

107. Moran L, Musselwhite E, Zenger JH. 1996. *Keeping teams on track*. McGraw-Hill, New York.

108. Hackman JR. 2002. *Leading teams: Setting the stage for great performances*. Harvard Business Review Press, Boston.

109. Grant A. 2023. *Hidden potential*. Viking, New York.

110. Bryant A. 2011. *The corner office: Indispensable and unexpected lessons from CEOs on how to lead and succeed*. Times Books, Henry Holt & Co, New York.

111. Search filings. U.S. Securities and Exchange Commission. https://www.sec.gov/search-filings [accessed December 31, 2025].

112. Yahoo! Finance. https://finance.yahoo.com/ [accessed December 31, 2025].

113. Dennison K. 2023. How ghost job postings are creating a false sense of hope. *Forbes*, November 27. https://www.forbes.com/sites/karadennison/2023/11/27/how-ghost-job-postings-are-creating-a-false-sense-of-hope/.

Index

R

S

T

About the Author

Carl M. Cohen, Ph.D., is President of Science Management Associates (www.sciencema.com) and provides coaching, consultation, and training in interpersonal, group, and organizational skills to scientists and science executives in both the public and private sectors. Carl is widely recognized as a global thought leader in best practices for managing and leading scientists and scientific organizations. He has more than thirty years of biomedical research and management expertise, including as a senior executive in the biotechnology industry and as Professor of Medicine and Professor of Anatomy and Cellular Biology at Tufts University School of Medicine. Carl is coauthor, with his wife, Suzanne L. Cohen, EdD, of *Lab Dynamics: Management and Leadership Skills for Scientists*, 3rd ed. (Cold Spring Harbor Laboratory Press, 2018). Carl has done training and on-site management consulting for universities, top ten pharmaceutical companies, and biotechnology companies both in the United States and internationally. Over the course of his career, he has coached and mentored hundreds of working scientists, science executives, and leaders of academic and private research organizations. He was the founding director of and ran the Cold Spring Harbor Laboratory Workshop "Leadership in Bioscience" from 2011 to 2019.

www.ingramcontent.com/pod-product-compliance
Ingram Content Group UK Ltd.
Pitfield, Milton Keynes, MK11 3LW, UK
UKHW060220061026
15625UKWH00007B/418

9 781621 825555